MEDIA, POLITICS, AND DEMOCRACY

USED BOOK

W79

F

USED PRICE

Sell USED Buy USED AT THE MSU BOOK STORE

$ 3.70

RECONSTRUCTION OF SOCIETY SERIES

General Editors
Robert E. Lana
Ralph L. Rosnow

AGGRESSION AND CRIMES OF VIOLENCE
by Jeffrey H. Goldstein

THE END OF IMPRISONMENT
by Robert Sommer

MEDIA, POLITICS, AND DEMOCRACY
by Bernard Rubin

MEDIA, POLITICS, AND DEMOCRACY

BERNARD RUBIN

School of Public Communication
and The Graduate School
Boston University

New York
OXFORD UNIVERSITY PRESS
1977

Copyright © 1977 by Oxford University Press, Inc.
Library of Congress Catalogue Card Number: 75–25461
Printed in the United States of America

for

Ruth,
Roberta Louise,
our families

INTRODUCTION TO THE SERIES

When society requires to be rebuilt, there is no use in attempting to rebuild it on the old plan.

—John Stuart Mill, 1858

The problems of society are many and complex, and sound plans for revitalizing it are few. Institutions that once seemed to work quite well no longer do so; and many new programs, hastily conceived, have failed or just faded into oblivion. Therefore, it is time now to determine what has and has not been accomplished, to analyze and challenge our assumptions, and to offer new and carefully conceived blueprints for rebuilding society. At the same time that we examine the problems of society, we must also consider the institutions that function as agents of change.

That is the purpose of this book and the others in the Reconstruction of Society series. Knowledgeable social scientists have been asked to direct their expertise toward solving some of the problems of society—to develop plans that are specific enough to form the basis of policy discussions and decisions. In the end, of course, even the best blueprint can be no more than a technical exercise if the means and the will to bring it to fruition are lacking. Justice Felix Frankfurter once wrote: "In a democratic society such as ours, relief must come through an aroused popular conscience that sears the conscience of the people's representatives." The ideas presented in this series of books may be a step in that direction.

The mass media are as much the people's representatives as are the individuals elected to reflect the popular will. Both are sounding boards, both help to fashion a picture of reality, of what is important and unimportant, and both are essentially involved in the processes of social change. The mass media (the electronic media especially), however, may be caught in a double bind. On one hand, treatment of ideas in the extreme may "condition" the public to a superficial level of understanding. On the other, when the media stand above a problem and judge it impartially, the public may come to see itself as a neutral spectator of partisan conflict. Add to this that public access to the mass media is greatly restricted, and the issue of social responsibility becomes increasingly complex. Professor Bernard Rubin, in this third volume in the Reconstruction of Society series, elucidates these complicated issues with penetrating analyses and practical remedies. The mass media have a profound responsibility: they must be sensitive and responsive to societal problems while endeavoring to avoid creating problems themselves.

Philadelphia, Pennsylvania Robert E. Lana
June 1976 Ralph L. Rosnow

CONTENTS

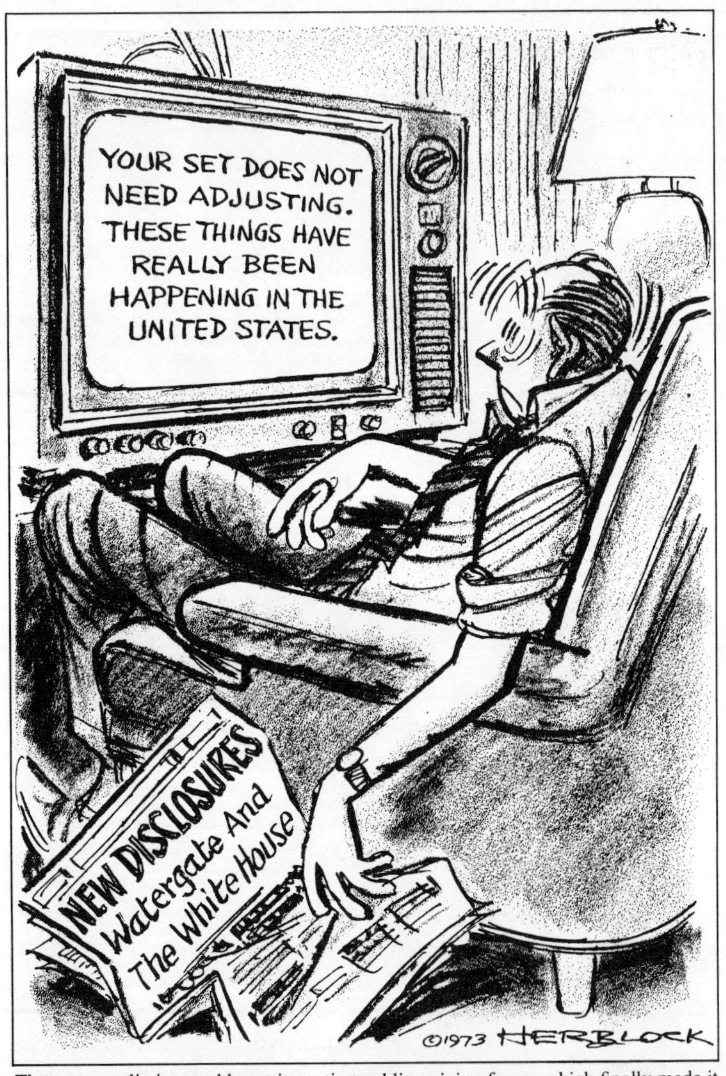

The mass media inexorably set in motion public opinion forces which finally made it impossible for President Nixon to retain his power. Nevertheless, the Watergate story offers students of politics little evidence of planning or premeditation by media leaders. While a few encouraged investigative reporting and full analysis, most merely followed the outline of the scandals slowly emerging. It is obvious that in regard to most of the important subjects of our time, the media are not yet organized to deal in depth with issues or to research for hard-to-get facts.

PREFACE

The mass media, for better or worse, are the most dynamic agents of change in today's world. Basic social and political education is so strongly influenced by what is presented that the media must be appreciated as elemental in communication within all public schooling systems.

This book deals with the ways in which the media handle political events and processes. Implicit throughout the discussion is the premise that the survival of democracy depends upon the success of those critics and practitioners who are determined to manage media in the public interest.

Advocates of freedom are well aware that legitimate defenses of rights and interests can be subverted if the people do not understand the relation between key events and vital political processes. Unscrupulous power brokers have often benefited from failure of the media to alert the citizenry to dangerous situations. It is all too true that the educational power of the media has frequently been less obvious than the tendency to disseminate superficial or misleading information.

In order to emphasize the media environment that itself creates political problems, five key topics are analyzed in some detail:

communications objectivity, community value setting, media freedom, popular participation, and election trends.

Chapter 1 deals with the connection between mass media objectivity and public understanding of political reality. Significant recent cases are discussed, and related communications trends analyzed.

We need more media work dealing with the range of alternatives facing decision makers and the public they serve. Rarely are decisions connected to the events or circumstances that preceded, and new problems are not often traced to vital decisions that set off waves of effects on society. In short, the connection between plans and results must become a more routine aspect of media presentation. With the goals of society being increasingly set by governmental institutions and private interest groups, the media can make its greatest contribution by encouraging meaningful popular participation in government and in private organizations benefiting communities. To that end, Chapter 2 describes the connection between social values and the programming alternatives confronting media managers who desire to stimulate such participation.

Intervention as a communications concept is treated theoretically and then applied to specific examples of ways in which the media have been influential.

Since rational behavior is vital to legitimate government and to social well-being, special attention is given to the media's obligations to help reduce violence in everyday American life. Ironically, such common violence has attracted less than adequate attention, while many people have concentrated on brutalities connected with international wars. We must concern ourselves with both problems. One of the goals of this volume is to show how violence experienced by whole classes of the citizenry has been ignored by the media and therefore by their vast public. Another is to show that the mass media have glorified certain types of violence, particularly in entertainment programs, and thereby detracted from the efforts of responsible citizens working for the abatement of violence in society.

Equally important is the struggle to maintain and enhance press freedom, and this subject has been treated at some length in this book. The constitutional guarantees of freedom of the press are only as valid as the determination of each new generation of Americans to defend the law in letter and spirit. In recent decades there have been occasions when it appeared possible that intimidation from private and government sources might be more than the press's champions could handle. When national security concerns, both legitimate and imagined, have dominated the news, it has become necessary to defend free inquiry against a series of highly emotional attacks.

This post-Watergate, post-Vietnam era is a fitting time to assess the contest between proponents and opponents of the concept of free media, and to appreciate just how close that contest has been. Chapter 3 does not offer a lopsided description, with a courageous press on the one side struggling with extremely powerful outside elements. On the contrary, it points out that not all media professionals have demonstrated the courage or the competence necessary to the survival of free media and, by extension, of democracy itself. One concludes that recent history should be carefully analyzed by all citizens worried that the defenders of free press traditions could lose upcoming struggles.

The mass media do not constitute an institution to be monopolized by a small minority indifferent to the needs and opinions of society as a whole. On the contrary, the quality and degree of access made possible for the individuals and groups in our society determine how democratic the media are.

Several of the most important aspects and problems regarding access to the media are discussed in Chapter 4, with illustrative cases. Demands for participation in media organizations and for adequate media coverage of minority interests are shown to be crucial.

Of equal importance is the subject of censorship, and Chapter 4 presents specific commentary and examples bearing upon the difficulties faced by honest reporters who want to get stories. Special attention is devoted to the psychological bases of repression of the press, and governmental attempts to stifle press inquiries are illustrated. The national security blanket has too often covered situations that should be brought before the public. Habits of public officials grown used to withholding information, with little or no justification, are traced to emotions and attitudes connected with national crises. We all would do well to reflect on how to avoid the replay of yesterday's news, with the lessons of the recent past in mind.

Chapter 5 concentrates on current problems of media politics as they relate to elections. Important current developments are discussed, with emphasis on reforms enacted or proposed to control the money that pours into political campaigns and to make the new breed of professional political consultants accountable to the politicians for whom they work. Without such reforms, elections would become less necessary as power gravitates to those who know how to manipulate the media and thereby to dominate politics. Few long-term solutions are offered, because the reforms following the Watergate scandals have not yet been tested. By the conclusion of the national election campaign of 1976, we shall know a good deal more and be able to forecast on the basis of new developments.

Media politics is relevant to every individual and group in this nation, since every special interest is dependent, to a substantial degree, upon the adequacy of the mass media. Where political matters are concerned, the print press, radio, and television are obliged to be fair and competent. Beyond that, the general public has a right to expect an ever-increasing amount of work that can be rated excellent. Excellence, when it appears, is the product of more than basics: creativity, curiosity, and courage are vital contributing characteristics. The public must demand excellence from mass media entrepreneurs if democracy is to have a future.

As I think about alternatives for our future society, I recall vividly the many events of our recent past which the media made stark and personally real to me. What I read about those events, supplemented by what was conveyed by radio and television, has compelled my concentration on basic problems of political communication. Looking back at some of the more tragic stories forces a search for ways to spare upcoming generations the necessity of repeating follies.

I would follow the advice of the Athenian Stranger who in Plato's *Laws* says, "Speaking generally, our glory is to follow the better and improve the inferior, which is susceptible of improvement, in the best manner possible." * Trying to do so, all the good works of the media give me hope, even though there is such waste of social opportunity by its managers. For all of the players' frivolities with the instruments that do shape lives, there are many achievements. One achievement is that the media sometimes burn images into our minds, images to which we react in countless ways for years afterward. This book is one product of much thinking stimulated by events like those of a very short period in 1975.

April 29, 1975, was a cold day in New England. Winter forgot the calendar exits planned for the seasons and grasped at every emerging bud. Those harbingers of Spring revival—the crocus, daffodil, and tulip—normally flowering in sequence, pushed forth toward the weak sunlight together. It was a rugged ending, not completely surprising, that was very soon to be swept into memory by bright rays of the sun.

In Saigon it was naturally tropical and, courtesy of mankind, terrible. The last remnants of organized resistance to the advancing North Vietnamese and Viet Cong forces were crumbling. More than a month of rapid change had seen the disorganized retreat of President Thieu's best armed forces units turn into complete rout.

Here in the United States, newspapers, radio, and television, that day as usual, mixed entertainment with enlightenment, making sure that their audiences were satisfied. Incredibly, even the death throes of South Vietnam had been taken in stride by our people reading about, listening to, or watching the media accounts of agony and anguish half a world away.

For a short time, during the preceding days, there was a national

---

* B. Jowett, ed., *The Dialogues of Plato,* Vol. III (New York: National Library Company, n.d.), p. 426.

outburst of sympathy for Vietnam orphans airlifted to safe haven in the United States. Against the backdrop of the news about the climactic events in Cambodia and Vietnam, which had been ruined by big-power politics, the "save-the-babies" stories aroused hope in so many hearts that a semblance of humanity had been salvaged from the carnage.

That night the television networks, on regularly scheduled news programs and on *specials,* colored in the stories that filled newspaper headlines and columns. President Ford, informed that Saigon would fall to the invading army at any moment, ordered the evacuation of the last Americans and of Vietnamese who had committed themselves to our government's efforts in that country. By the miracle of satellite television we were able to peer inside the United States Embassy compound in Saigon as the Marines rushed in troopers to protect and helicopter out refugees. For nineteen hours and more, the "choppers" flew in from fleet units to ferry out the desperate from a helipad improvised on the Embassy roof.

We were there, courtesy of freshly filmed scenes and *live* event coverage, as they raced to the escape, little more than seventeen minutes away and just beyond the South Vietnamese shoreline. We were there, at the Embassy compound gates as Marines and armed civilians prevented screaming, frantic crowds from entry. The gates were occasionally opened, in a weird game of roulette with lives, and for obscure reasons. Some persons were let in and many were thrown back. People pleaded to be allowed to squeeze through, offering money, excuses, heart-rending (and valid) reasons to the guards.

Every time the gates were pulled open a little to let a favored one or two in, others screamed and hurled themselves against the grotesque scrimmage line. Some got through, clenching children or even heavy hand luggage. We television viewers observed the rules of the game! Once through the barriers you were safely en route to the helicopters, even if no one had said *Enter.* With the rough justice enforced at the gates, the rules provided that some people who attempted to climb the Embassy walls and deal with flesh-tearing horrors of the barbed wire, had their fingers clubbed as they neared the top and their heads bashed if they attained it.

Hours later it was ended as the last "chopper" departed and the Embassy was open to the looters and sackers who roamed the streets. The victors took over the city, and an era ended. There was no light for the United States at the end of the tunnel. There was only the rockets' glare from the advancing army's bombardment to light convulsions signaling *finis* to thirty years of conflict, including more than ten years of active American involvement.

In the next days and weeks all segments of our press devoted considerable time and space to reportage of other connected stories and to recapitulation of the background, which spanned decades. Somehow, to this observer the traumatic scenes at the Embassy were as awesome as any of the war. They have burned themselves deep into heart and mind.

A few days later the media were reporting discontent in certain American communities where citizens protested against the entry into this country of tens of thousands of Vietnamese refugees, when so many people were out of work. Others countered with equal emphasis that the gates should be opened: it was the least we could do, after all that had happened, they said.

This is a book about mass-media reform based on the premise that we must all understand the substance and procedures of mass communication, if a turning away from chaos and terror is to become evident in the world.

This is a book about reform, about social construction and public harmony. This is a hard look at the suggestions that have been made for improving *the system,* emphasizing the roles of the mass media.

This is a book that recognizes the power of events and the importance of value judgments about them.

Open the gates. Let us all flee from disaster!

Boston, Massachusetts Bernard Rubin
April 1976

COMMUNICATIONS OBJECTIVITY AND POLITICAL REALITY

1

THE SITUATION

Contemporary life is dominated by technology. The development of media for mass communication constitutes potentially one of the most constructive contributions of technology insofar as democratic community goals are concerned. The broadest possible dissemination of knowledge has always been regarded as the key to a well-informed citizenry. That goal notwithstanding, there is always to be considered the possibility of the means of this dissemination being controlled by unethical, immoral, and ill-trained managers. Accompanying the possibility of universal intellectual advance is the possibility of half-truths, partial answers, and unimaginative solutions.

When the public seeks deeper political understandings, simplistic views are too often provided. Even basic national and international problems are presented as entertainment. Media performance seesaws, at times deserving high praise, at times deserving censure.

The mass media are involved in the political life of the nation to an extent never envisaged by the founding fathers. For most citizens, conceptualizations of political issues, events, and personalities depend upon the mental images drawn by editors, publishers, reporters, commentators, columnists, and professional critics. The media are so vital to most citizens that any political authority which could deflect members of the media from professional objectivity could also begin the process of destroying this democratic society.

SAFEGUARDING THE PUBLIC INTEREST

Despite the understandable reluctance of leaders of the mass media to subject their enterprises to the scrutiny of any independent source of regulation, there is increasing public pressure for objective reviews of the press. It is felt that biased and distorted reports on public affairs would be less numerous if standards of professional work were raised and were applied on an industry-wide basis. The Fourth Estate, according to the view held by the majority of supporters of our Constitution, must be free, bold, *and responsible*. Reporters, for one example, should not deliberately produce news accounts showing only one side of a complicated and controversial situation, and thus tilting public opinion irresponsibly. For another, there must be constant protection of the civil rights of all those whose actions are reported. The First Amendment to the Constitution orders that "Congress shall make no law . . . abridging the freedom of speech or of the press." The press must enhance those freedoms by giving the American people the fairest and most honest reports possible.*

---

* Disputes are constantly triggered by press reports and editorial positions. Two recent examples illustrating strained relations between business and the press follow. WNBC in New York City ran a five-part evening news series on the background to gasoline prices. The leaders of Mobil Oil Company thereupon placed a full-page advertisement in the *New York Times* (Mar. 5, 1976) to condemn what they labeled "hatchet job" work by reporter Liz Trotta. The Mobil response offered *facts* to counter eighteen of her statements. Mobil titled the advertisement "What ever happened to fair play?" and called the reports "inaccurate, unfair, and a disservice to the people."

Similarly, the American Electric Power Company, Inc., paid for a long countering advertisement to fight press opposition to its proposed Blue Ridge hydroelectric power project on the New River, which flows through West Virginia, Virginia, and North Carolina. In particular, a *New York Times* editorial had advocated that the

Post-auditing the Press: The National News Council Approach

On August 1, 1973, a private organization, financed primarily by grants from philanthropic foundations and press organs, began operations. The National News Council (NNC) was dedicated "to serve the public interest in preserving freedom of communication and advancing accurate and fair reporting of news."

Its fifteen members, five advisers, and staff together represent a very high level of experience and leadership drawn from the fields of law, education, mass media, civil rights, and politics. As of this writing, the chairman is a former Chief Justice of the Court of Appeals of the State of New York, and the vice chairman is director of the Aspen Institute's Program on Justice, Society, and the Individual. The membership includes the creator of "Sesame Street" and "The Electric Company" for public television; a former editorial-page editor for the *St. Louis Post Dispatch;* a distinguished former congresswoman from Oregon; a Methodist pastor who is a leader in the nonviolent civil rights movement; and the managing editor of the *Chicago Sun-Times,* who has been president of Sigma Delta Chi, the society of professional journalists.

Secretary of the Interior rule that the river be considered "wild and scenic" and protected against the project by "bureaucratic resolve." American Electric pronounced "this unbalanced journalistic presentation outrageous" and satirically noted, in a reference to the famous motto of the newspaper, that "the whole truth—in case some have forgotten—is fit to print" (*New York Times,* Jan. 9, 1976).

Who is to judge in such disputes? Are possible remedies forthcoming from sources other than the law courts and regulatory bodies such as the Federal Communications Commission? A large and probably majority segment of American news editors, publishers, reporters, and producers is chilled by the very idea of further possibilities for intervention, mandatory or optional. Many of those who feel that way consider themselves *absolutists* on the First Amendment, regarding any small measure of additional regulation as a step taken on the road to an unfree press. Hence their criticism of organizations such as the National News Council is quite understandable. One well-known critic of NNC is Nat Hentoff, the investigative reporter who delves into highly controversial issues. He considers himself an absolutist on the First Amendment and frowns on such government-imposed guidelines as the "Fairness Doctrine," which applies to electronic media but not to the print press: "There can be no concrete proof of how much bolder and braver television journalism may become if it is fully protected by the First Amendment, but surely it's an option worth taking." (See Nat Hentoff, *How Fair Should TV Be?,* Television Information Office, New York, 1974, 8 pp.)

Patterned on older, more established press councils in England, Norway, Sweden, Denmark, the Netherlands, Austria, and West Germany, the National News Council attempts to review the work of "national news suppliers." It encourages the development of other press councils at local, state, and regional levels of activity.

Surveying its area of interest, the Council finds no shortage of concerns, including:

> . . . full disclosure of possible conflict of interest, access, accountability, the payment of fees to sources of information, or "checkbook journalism"; the effect of monopoly ownership on press freedom, the subpoena of reporters' notes and out-takes of materials produced for radio and television, shield laws to protect against such subpoenas; the imposition of gag rules in the judicial process, with its concomitant fair trial/free press controversies; deliberate editorial distortion, bias, and, of course, accuracy and fairness.

The *New York Times* has not been in favor of the Council, and from the start has withheld significant cooperation by denying the Council's requests for information relative to public complaints against the newspaper.

Another jurisdictional problem relates to the Council's authority in regard to electronic journalism. A task force that originally studied the Council's "feasibility" recommended that its "processes, findings and conclusions should not be used in F.C.C. [Federal Communications Commission] proceedings." The Council is trying to develop appropriate procedures to conduct its work in the field of electronic journalism.

From August 1, 1973, to July 31, 1975, the Council reviewed and publicized its decisions in 59 cases in which complaints were raised against the media. Relatively few of those complaints, 5 to be exact, were upheld, and 33 were declared to be unwarranted. Some 21 complaints were dismissed because the Council found that the complaining persons or organizations did not "pursue with specifics" or because the complaints were "beyond the Council's purview."

The greatest number of complaints, twenty-four in all, were

directed against television networks. As of the end of July 1975, two of the complaints had been upheld. Of eleven complaints concerning national newspaper stories, one, involving the *New York Times,* was upheld. Another complaint was filed by John Haydon, former governor of American Samoa, against NBC-TV on November 11, 1974. He criticized the "Weekend" program about that country as inaccurate and "designed deliberately to malign the Samoan people, the administration of the territory, the Department of the Interior."

NBC decided to stay away from the Council's hearing on several grounds. One was the determination that if NBC-TV had to defend itself before any forum, it would be the Federal Communications Commission, to which it is legally accountable as a licensee. The network did cooperate with the Council to the extent of providing transcripts of the program under scrutiny and arranging for a viewing of a taped recording.

A number of experts on Samoa testified. Included were an educator and a television education specialist, each with about six years of work experience in Samoa behind him, former government officials, professors of anthropology, and a television producer who created a program about the territory in 1967. Dr. Margaret Mead gave oral testimony, as did an expert from the Department of the Interior who had visited Samoa frequently in the course of his official duties. Learned studies, government reports, and press clippings were carefully examined.

The Council decided that the accuracy of the program on Samoa was not up to professional standards. Among its conclusions:

> . . . While great latitude must be accorded to television producers in the case of any given documentary, that is not to say that there is not, or ought not to be, a limit to the degree of distortion and misrepresentation that a producer can indulge in. We believe that the NBC documentary on Samoa clearly exceeds that limit.
>
> . . . The aforesaid distortions and misrepresentations go well beyond any that could be justified under the rubric of robust journalism, and to that extent we find the complaint justified.

In another controversy over accuracy, the Council found a complaint against the *New York Times* justified on June 25, 1974. The case involved an article about an important scientific study of the effects of herbicidal spraying in South Vietnam. Dr. Anton Lang, a member of the National Academy of Sciences, which sponsored the study, complained that the *New York Times* and the *Washington Post* published an article by John Finney (on February 22, 1974) "prior to the official release of the findings of the study." Dr. Lang charged that the information reported "[was] based on a leak, contained outright errors, was slanted, disregarded important constructive aspects of the report." (Because the New York Times News Service "moved" the article in question, the Washington Post was seen by the Council as outside its purview.)

The Council found that the article "relied heavily on information obtained from those who disagreed in part with the majority conclusions." Therefore, "the Times had a special obligation to its readers, upon official release of the report, to disclose any differences between its original article and the official version." Further criticism was leveled at the newspaper for its failure to print the text of a letter of complaint sent to it by Dr. Lang and its failure to state, in a fuller article later giving more details on the majority consensus of the scientists who participated in the study, "that the original article was incomplete." [1]

The NNC has served the press well during the short period it has been in existence. It is clear that its primary effort is to influence rather than to regulate media professionals. The Council may be of great moral influence upon this generation of journalists who hear of its work and are aware of activities that are frowned upon.

MEDIA GUIDELINES FROM GOVERNMENT: DO THEY GAG THE PRESS?

Since the Nixon administration's involvement in the Watergate affair, the press has been more suspicious than ever of all governmental attempts to guide the media and to keep information secret. Its fear of political direction has been further aggravated by the memory of all the misinterpretations and inaccurate information supplied by government information officers during the entire period of American military involvement in Vietnam.

A recent example of governmental control over what the media may and may not disseminate deserves our attention because it illustrates how very sensitive reporters have become.

U.S. Supreme Court Justice Harry A. Blackmun issued a ruling relative to a Nebraska criminal case on November 20, 1975. (He was acting in his capacity as the high court's justice assigned to deal with cases in the judicial circuit containing Nebraska.)

The crimes being dealt with in the Nebraska court were the murders of six members of a family on October 18. Allegedly, the man accused of the murders made statements to local law enforcement officers, among which was his confession. The lawyers for the prosecution and the defense requested a "protective order" restricting the information the media could report, and a county court judge granted one. A higher state judge substituted his own protective order several days later, stating that it was "inappropriate" for the press to publicize the confession.

An appeal was filed with Justice Blackmun to rule against the order as a violation of the First Amendment of the Constitution. He ruled instead that, to ensure fair trials, state courts could prevent the press from reporting on certain information relating to the pretrial period of criminal cases. Certain statements by a defendant, confessions, and a defendant's previous criminal record were examples of what could be stopped.

One of the key rulings against such juridically enforced prior restraint of the press, popularly labeled "gag orders," was handed down in 1971 when the U.S. Supreme Court decided against the government's order forbidding publication of the Pentagon Papers, the classified government-prepared analysis of the nation's involvement in the Vietnam war. By a vote of 6 to 3 the Court declared that the government must prove that publication "will surely result in direct, immediate and irreparable damage to our nation or its people." Justice Blackmun dissented in that case, taking the view that publication of the Pentagon Papers "could clearly result in great harm to the nation,'' harm such as ''the death of soldiers, and destruction of alliances, the greatly increased difficulty of negotiation with our enemies, the inability of our diplomats to negotiate. . . .''

There was another aspect of the Nebraska case. Nebraska is only one of twenty-three states that have adopted voluntary press-ban guidelines as a means of avoiding publicity prejudicial to fair legal proceedings and avoiding arbitrary legal controls. Many media news professionals are leary of such guidelines, fearing that they could lead to habits which might be covered by law later on.

Blackmun's ruling was viewed by elements of the press as damaging to First Amendment guarantees on at least two counts. First, the press-rights issues debated in the Pentagon Papers case which revolved around matters of national security were cropping up in regard to criminal cases. Second, Blackmun's ruling was interpreted as opening the legal door for the courts to mandate guidelines for news about criminal trials.

Worry is widespread that there is not only a gag-order precedent but also a precedent for a new and possibly undemocratic press-government relationship.[2] Every responsible person is in favor of fair trials *and* national security. However, the essential question which must be answered if political life in this country is to remain democratic is: Should the crucial duties of the nation's press be defined and then regulated by government powers that be?

Justice Harold R. Medina, a senior judge of the U.S. Court of Appeals for the Second Circuit, in an unusual private action, objected to ''Omnibus 'Gag' Rulings,'' as he entitled his *New York Times* article of November 30, 1975. He pointed out that gag rulings worked against the press's public service of ''ferreting out hidden and obscure circumstances connected with the commission of crimes but more particularly in exposing official corruption and laxity by the very persons who are supposed to be hot on the trail of those who committed the crimes.'' Medina might have been thinking of criminal cases of the type that local and state law enforcement officers deal with every day. However, it does not stretch the imagination to believe that he might also have sensed connections to the sort of sordid activities and circumstances which, when finally revealed by the press, caused the resignation of President Nixon, among the other serious results. He was blunt in his advice:

> As I never did like censorship of any kind, I have a little unsolicited advice to the news media. First, I would stand squarely on the First Amendment itself. I used to think that guidelines might be helpful. Now I believe them to be a snare and a delusion. And the same is true of legislation. . . . Second, I would make no compromises and no concessions of any kind. Third, I say fight like tigers every inch of the way.[3]

THE FIGHT AGAINST POLITICAL APATHY: NEWS OR SENSATION?

Except for their sleeping hours, most Americans go through life as sound and information addicts, with news one of their great cravings. Psychologically, such people are of the greatest professional interest to social and political scientists, who are beginning to understand that the news so sought after may not be fundamentally important to the receivers as intelligence about events and developments. Indeed, the outpourings of news by the mass media may be significant primarily as building-block ingredients making up so many *life styles*. Unless news is directly perceived as signaling a potential or actual personal threat, most people accept such information in the same manner as they accept the sounds of the music they like. News is sensed rather than appreciated or analyzed. News has become so much *background stuff*.

If this phenomenal interpretation is valid, it explains why politicians, government administrators, and professionals of the press are finding it increasingly difficult to move the citizenry to action. It may be the important clue sought by journalists, teachers, and all others devoted to interpretation of metropolitan and cosmopolitan developments, as to why the mass media are ever more competent as disseminators of information while their audiences and readers are ever more complacent.

The news industry's productions may be elementally *sensational*.

In short, the sensory effects of news, the psychical impact on individuals, may steadily reduce interest in or concern for perception of what is happening in the world beyond that which is personally attractive or interesting.

That hypothesis may account for the ever-increasing tendency of the news media to treat political events as theatrical occurrences. Media managers might be adapting their professional work to what they perceive as important psychological requirements of the general populations they address. By playing up the bizarre or the dramatic, the more easily digested stories—political stories especially—those managers avoid intellectual confrontations with their audiences. When events are deemed newsworthy, certain literal facts are highly publicized. Normally, the superficialities are played up so as to appear fundamental to the stories.

Terrorists strike, but the moral and political arguments raised against their rampages are ignored or played down; the consumer price index is recorded at a new inflationary high, but the increased economic pressures on millions of people are less publicized than are the statements of callous bankers who declaim on calculations of the gross national product; a school busing crisis in a large city goes on and on, with street-level tensions steadily rising, and little cogent information is provided as to why the situation is out of control of professional educators and politicians; crime rates rise phenomenally, and news reports are largely bulletins about criminal acts, with little or no attention to victims.

One could misinterpret the results of all this catering by media to the supposed psychological needs of individuals and conclude that audiences are satisfied. The contrary is true! The general public is more suspicious of the press than ever before. Superficial reporting is a response to only one of its needs. The public is more and more alarmed by the news of politics and government and community life in the nation and around the world. People are more afraid, more prone to extremist thinking, more convinced that they are ignorant on too many subjects. Government is ever more distrusted, and the press is suspected of being part of a system dedicated to deceiving the public. That is the political environment in which we all have to live.

A few illustrations of this state of affairs were offered by the participants in an Educational Broadcasting Corporation "Behind the Lines" program televised in New York City on January 2, 1975. Host Harrison Salisbury, of the *New York Times,* discussed press problems with media professionals of the Boston area. Significantly, the program was entitled "Public Enemy Number One." Salisbury, in his introduction, said, "All across the country, people feel alienated from the press; they're suspicious of what they read in the newspapers and what they see on television. They're suspicious of the motives of reporters and editors, and they just don't see their world reflected in the press."

Alan Lupo, a free-lance reporter who has worked for newspapers and television stations in the Boston area, talked about one revealing situation involving a woman who lives in a housing project in South Boston. She called some press organizations and told them "Something to the effect," recounted Lupo, "that 'We've had it, we want to explain to you people what it is we're objecting to down here, why this project is so awful, and we'd like to get something done about it.' " Lupo said that at least two television station crews arrived and began to unpack their gear. Suddenly, from the portable radios which keep them in contact with their studios, a "Fire, fire" announcement was heard. They all packed up to dash to cover the fire story. The woman tried to get them to stay and pointed out that they were heading for an abandoned building. They went anyhow because, deduced Lupo, "television was saying to that woman . . . that a fire in an abandoned building was more important than what she had to say in that housing project. And I find that very dangerous."

Salisbury analyzed the coverage of the Boston-area press on school desegregation stories to that date. He appreciated the predicament reporters are in when they try "to strike a norm between violence and reason." He felt that the effort to do so had not lasted long. After pointing to the highly sensational press treatment given to the incidents of violence between blacks and whites, Salisbury declared that "Boston today is a city polarized, almost terrorized by the issue of busing. And it is a city polarized over the issue of press and television reporting. There is not much middle ground left." He somberly concluded, "About the only thing that both sides seem able to agree on, whatever their other views, is that the press has distorted the basic issues." [4]

ASPECTS OF CIVIC EDUCATION

Occasionally, but more infrequently than is healthy for a democracy, the news is made meaningful to the majority of citizens and they become personally involved in the play of events. The recent Watergate story alerted our society to the crucial matters before the nation: the required ethical standards for politicians, the future of the democratic system, the arbitrary powers vested in the President and in the executive office he heads, the rights and duties of Congress and the federal courts, and the obligations and privileges of individuals in society.

The Watergate story is important not only because a President of the United States was forced to resign, in large part because the media shook the American public out of its complacency for a time. It is important also because the mass media fulfilled all their essential political tasks. Henry Fairlie, an astute British observer, summarized those tasks:

> First, to try to reconcile the multiplicity of conflicting interests and will which exist in any free society and to produce from their conflict a policy. . . . [Second,] to maintain public interest in political issues, for without such interest free government is meaningless. [Third,] to act as a catalyst on public opinion. [Fourth,] to be a link between informed and public opinion. The two are very different.[5]

Coverage of the Watergate story by truly objective and professional newspeople showed the press at work on the moral background to politics. Bizarre Watergate developments highlighting the spying and lying by the conspirators and their dupes forced analyses of political morality to the foreground, so far as press attention was concerned. Carl J. Friedrich, an eminent political historian and theoretician who has spent a lifetime fighting all forms of authoritarianism, reminds us that politics and morals are "closely intertwined and Aristotle was right in treating them in close conjunction."

The mass media did not deliberately provide moral guidance as they helped explain events surrounding Watergate. Nevertheless, their reporters and analysts did not avoid the moral questions behind the legal and political tangles, behind all the literal facts. Press stalwarts of the electronic and print media did a very creditable job making clear that Watergate involved a conflict between devotees of fundamentally opposing sets of political principles, with the future of constitutional government at stake. The general public depended upon the media as it learned about dubious and thoroughly reprehensible activities of government. The presentations of facts and of opinions changed the mood of the nation and influenced attitudes around the world. Whatever the political goals are, socially consequential outcomes depend upon "modern mass communications, which significantly alter the political environment and hence call for adaptation." [6]

PUBLIC SERVICE GOALS

Analysts of the mass media habitually urge greater educational emphasis, with less priority given to entertainment. The Federal Communications Commission's so-called Blue Book (the color of the cover) of 1946 (officially, *The Public Service Responsibility of Broadcast Licensees*) established two major criteria of public service. One was "balance" in programming by broadcasting stations. In part, this meant that "local, live" programs were to be in "sufficient number" in relation to recorded or network programs. The other was presentation of a "reasonable number" of programs in areas of education, news, public issues, agricultural interest, labor interest, nonprofit civic and religious subjects, minority cultural tastes, and so forth.[7] With modifications, the spirit of that FCC regulatory approach still dominates official thinking, even if it is weak in official application.

The 1946 guide presupposed that local radio stations and the radio networks could easily follow the regulations by checking their schedules and increasing or decreasing different types of programming as needed. This assumption is questionable. First, the commercial dominance of the broadcasting industry created conditions which rewarded local and network owners handsomely for allowing the transfer of real power to advertising agencies and their clients. The civic-educational effort was, perforce, weak. Second, there was no real understanding of how broadcasters could ensure that their civic-educational programs would be significant presentations rather than mere showcases of whatever the quality of local leadership or talent was. Third, network presentations tended to be designed to be inoffensive to all elements of the nation's population and were therefore not usually of real social or political relevance to any one group. Television, the medium of communication which has become increasingly significant in American life since the early 1950s, was operated under the same general guidelines with equally poor results.

Print media have always surpassed electronic media in balancing education and entertainment. The plain fact is that newspapers, in particular, have always provided material of greater variety and searching civic significance than radio and television have, primarily because "balance" is rarely a major concern; newspapers can easily accommodate any and every type of story, from the frivolous to the most serious.

Another advantage of the print media is that their expenses are not determined by type of story. In the case of electronic mass media, however, expenses vary with each presentation and tend to be very high. For example, television talk shows originating in studios have lower production costs than interviews in the field. News documentaries are more expensive than summaries of events by news commentators. If television producers decide to tackle serious public affairs subjects, they must fight for extra funds to finance research and editorial services. Even in the new day of so-called instant communications, portable television cameras and mobile film crews require preplanned financial support far greater than that provided for journalists working on complicated stories for the print media.

In the United States radio broadcasting has been reduced to a doddering giant, when considered in terms of public service. Granting the usefulness of hourly news and weather reports and of talk shows, and granting the encouraging survival of some quality musical programming, radio has become more a producer of background sound than a source of foreground sense. Cost of production is not the only reason for the decline; lack of purpose, direction, and imagination is equally to blame.

Legions of great reporters work for newspapers and magazines, while radio and television can boast of only a small cadre. The electronic media tend to value personality and presentation style more than news savvy.

It is remarkable that the individual most revered by the television networks for dedication to news and public interest work is remembered as a unique specimen of talent. Edward R. Murrow, the quintessential reporter, made his mark in radio and worked hard to transport his ideas to television. "Hear It Now," which he created in collaboration with Fred W. Friendly, brought radio listeners the key events of each week through the use of sound recordings. For radio listeners Murrow's trenchant narration made news more alive than ever before. The first broadcast featured two U.S. Marine Corps enlisted men discussing their retreat from the Chongin Reservoir during the Korean war.

"See It Now" was first telecast on November 18, 1951. Murrow and Friendly sought to:

> . . . employ television as a primary news medium, rather than a secondary or derivative one. . . . *See It Now* was consistently ahead of the news, whether in its unoptimistic appraisal of the Supreme Court's 1954 school desegregation decision, its anticipation of the Twenty-fifth Amendment on the presidential succession, in its two one-hour programs on the emerging nationalisms of the African continent, or its two programs on cigarettes and lung cancer, years before the Surgeon General's report on the subject.[8]

Murrow's biographer, Alexander Kendrick, started in newspaper work but spent most of his professional life in television journalism as one of the "Murrow boys" at CBS. He knows full well that "See It Now" did not become a standard of the industry. Political life, Kendrick feels, has been much abused by television. He illustrates, decrying the use of "Cameras . . . not only to photograph urban rioting but often to encourage it; not to illuminate controversy but to exacerbate it, by their very presence contributing an element to the situation which altered its complexion and consequences." [9]

POLITICAL CHANGE AND MEDIA MANAGEMENT

Castigations of one or all of the mass media, stimulated by failures of these potentially invaluable instruments of education to respond constructively to social and political needs, should be based on appreciation of changes in elite group attitudes. After all, leading politicians, political scholars, and professionals in the mass communications field have come to understand only recently that ever-present social pressures are more important than election procedures and ideological statements.

A most interesting interpretation dealing with political changes in the last quarter-century concerns misconceptions held by political scientists themselves in their search for ways "Toward a More Responsible Two-Party System," as Evron M. Kirkpatrick, the distinguished executive director of the American Political Science Association, titled a 1971 periodical article. In it he describes misconceptions he shared with his colleagues while serving on the Association's Committee on Political Parties between 1946 and 1950. He directs scholarly attention to the failures of political parties; to the overesteem in which they were held by traditionalists who believed that all problems could be worked out rationally; to insufficient political attention to interests derived from social identities. "The electorate," he observes, "was conceived as consisting of discreet, thinking individuals, not persons bound into webs of identifications—ethnic, religious, economic, sectional, political—that condition individual political behavior."

Kirkpatrick suggests that the "Committee's premises were those of 19th century English Radicalism, garnished with touches of Rousseau." While reviewing certain aspects of this analysis, which explains his conclusion that the basic approach of his committee was wrong, it is well that we reflect on how much the mass media fell into the same traps in the last quarter of a century.

> Lack of concern with the social and cultural environment is one of the most striking aspects of the report. . . . Unconcern with the social, cultural, or even idelogical bases of its proposed party system [is another].[10]

> [He criticizes] (1) poor handling of value analysis, that is, inadequacies in goal thinking and derivational thinking; (2) failures to explore the interaction of the decisional system and the distribution of values, that is, of policy outcomes; (3) lack of sufficient attention to kinds of thinking peripheral to scientific thinking; goal and derivational thinking, trend thinking and developmental analysis; (4) insufficient empirical theory and inadequate data. . . .[11]

> The nonrational bases of political behavior are either disapproved and dismissed or ignored. An interest is not conceived as deriving from social identity, as naturally and legitimately influencing political action, but as an extrinsic, undesirable, and probably immoral interference with the public interest, which is nowhere defined.[12]

If political scientists were that far afield, in the view of an astute political analyst who is trying to explain why the general public has become so wary of traditional politics, what must we learn if the mass media are to perform more effectively? What would produce investigative reporting which would lead to better public understanding of and participation in political processes? What would newspapers, television, and radio have to be concerned with to deal adequately with politics in the public interest?

Obviously, media managers must be as courageous as Kirkpatrick and admit to "inadequacies in goal thinking." Indeed, his summary of the failures of analysis about political parties is, without change, pertinent to mass communications students who desire more media work in the public interest.

The mass media must concentrate on what can be termed the *anatomical* and *biological* factors of community life (local, national, international), in order to document the uneven progresses and regressions of civilization. Day by day the vast audiences must be offered information about the pressing issues which effect the quality of life.

Children and the aged, tenants in luxury apartments and ghetto dwellers, the governors and the governed, grade school strivers and collegiate sophisticates—in a democratic society, all need to know what is actually happening. Their knowledge needs to go far beyond the material presented in the superficial media reports which have preoccupied readers, listeners, and viewers. They need to know the details of serious and intelligent plans proposed to help move society away from disasters. They need to know enough to question leaders who decide what can or cannot be done. Such knowledge is necessary if citizens are to give effective support to honest and moral entrepreneurs and fend off demagogues.

Critics of today's mass media rightfully charge that they are "ritualistic in always reporting the same kinds of stories (events rather than situations)"; they are prone to stressing the "status quo and the establishment way of doing things" and are notable for a "penchant for sensation, emotion, pictures and action—a show-biz type of news reporting." [13]

Clifton Daniel of the *New York Times* spoke directly to the point: "Stripped to essentials, the responsibility of the reporter and editor is simply to serve the public—not the profession of journalism, not a particular newspaper, not a political party, not the Government, but the public." [14]

Specifically, how are the media to serve the public better and thereby directly foster increased political awareness? Conclusions from recent basic studies of society-wide problems suggest examples of poor performance by the media. The *environment* looms larger in the public mind. Failure of highly technological societies to curb industrial pollution of air and water is becoming shockingly evident, and failure to deal with the problem of overpopulation has already doomed a substantial portion of mankind to lives in which malnutrition is considered normal. A large minority of the world's population is starving to an early death, with no appreciable counteraction from places where gluttony is considered merely a health problem rather than an abnormal way of life.

There is a clear warning that natural energy resources on this planet are very limited. The oil shortages imposed by Middle East rulers in late 1973 actually resulted in discomfort in homes and factories in the industrial nations, which are considered advanced.

Great motorized monsters, weighing two tons and more and capable of only nine miles to a gallon, transported their owners to gasoline filling stations which doled out minuscule amounts of fuel at certain hours on certain days. Prices soared, but the real dimensions of the environmental crisis failed to reach most citizens. One concludes that a regional war and a worldwide political crisis were necessary to stimulate even rudimentary concern in the owners of those automobiles with the 300 or more "horses" under the hood.

Certainly, the true situations on the subjects under the umbrella words *environment* and *ecology* have not been systematically included in mass education. It is an awful state of affairs when such a basic and continuing crisis is apprehended more through individual sensory mechanisms than through mass media, a vital component of the social information system. According to Donald C. Pirages and Paul R. Ehrlich, in their treatise *Ark II,* this informational channel "in the past has not dealt much with environmental deterioration or the host of other issues that could lead to large-scale social change." They emphasize that "the dominant social paradigm shapes this information flow, and most of the people who influence media operations have little interest in raising potentially embarrassing questions about the present social, economic, and political system." [15]

Pirages and Ehrlich assume that basic political reformation will be required to change old habits of action and thinking. They suggest reallocation of values, changes in the methods by which leaders are selected, moderation of competitiveness in politics to instill a necessary level of cooperation, and encouragement of a new and highly developed social conscience. Also needed, they say, is a realistic review of the alleged benefits and limitations of reliance upon "political checks and balances," which "were designed to discourage hasty unreasoned actions, not to facilitate the transformation of a society headed for severe difficulties." In their view, "Politics is now dedicated to satisfying individual appetites, but is little concerned with future possibilities in meeting such desires." [16]

© 1973 by The New York Times Company. Reprinted by permission.

MEDIA AND COMMUNITY VALUES

2

What are our media goals to be if large segments of the population, now misdirected, ignored, or disadvantaged by mass communications, are to be socially benefited?

INTERVENTION THROUGH PLANNED INFORMATION

Democratically minded media managers want to encourage the creation of communications that will help people to lead more socially rewarding lives. Every one of their principal actions taken to help achieve that goal can be considered to be a political intervention, shaping the course of government and society.

According to a leading scholar on intervention theory as it applies to international affairs, "intervention refers to organized and systematic activities across organized boundries aimed at affecting the political authority structures of the target." [1] Within our nation, this concept of intervention applies to the works of one of the most active of all civic institutions, the organized mass communications industry.

On a daily basis, it exerts the most powerful single influence on the majority of the population. Without exaggeration, the media are largely responsible for how people think and feel about important issues. When, because of mutual interest or by accident, the mass media act in concert to bring a topic before the nation, that topic automatically gains importance regardless of its inherent significance. When the media reduce the importance of a topic by a casual and low-keyed treatment, popular interest is not aroused. Most of the time, mass media discriminate against topics that should be of popular concern by failing to plan at all for adequate presentation. Failure to use mass media constructively leads to increasingly greater conformity. Stimulation of individual thinking about crucial political questions is needed if democracy is to endure.

Media managers in radio and television are influenced by advertisers who are generally fearful of sponsoring programs that have political significance. The power of the advertisers is reflected in the labeling of the 1974 television year as the $2.2 billion TV season.''[2] As recently as 1967, national corporations manufacturing mass consumption products paid ''609 million . . . to control most of television content, incuding 95 percent of all prime-time access to America's households.'' Seventy-eight percent of the network's advertising came from ten giant corporations. The astute observer Ben H. Bagdikian summarizes the outcome of this massive corporate involvement: ''A whole generation's cultural values have sprung from that access. . . . For the most part they [the corporations] have not been interested in selling political ideas or ideologies.'' Also, he points out that in the process of pushing sales of products in every possible manner, the corporations ''unwittingly propagated social ideas like the glorification of violence on their entertainment programs in order to hold attention.'' In this celebration of materialism, most major social and political subjects are avoided.[3]

Among other powerful influences besides those exerted by corporations is the power of the President of the United States to demand access to a national television audience whenever he deems it in the national interest. Network news directors, especially during the years of the Nixon administration, learned how hard it was to resist a President who, along with his Vice President and key aides, used every tactic to try to convince the American people that the television networks were biased in their coverage of the Vietnam war and other key events. They became highly sensitive and were fearful lest Mr. Nixon provoke administrative actions against their organizations. After Vice President Agnew denounced what he called "instant analysis," the networks shied away from presentation of commentary about a Presidential speech on the same day it was delivered.

Newspapers of high quality—the *New York Times, Wall Street Journal, Washington Post, Christian Science Monitor, Kansas City Star, St. Louis Post Dispatch,* and others—manage, on a daily basis, to provide their readers with information of great educational importance. For the alert and concerned citizen there are also first-rate magazines such as *The Atlantic Monthly, The New Republic, Harper's, Trans-Action, Newsweek, Time,* and *U.S. News and World Report.* In contrast, coverage of news and public affairs by local radio and television stations and by the networks is superficial. The print media can be much criticized, but the dominant organs are not yet the handmaidens of the major advertisers.

A NEW CONCEPT NEEDED TO REFORM MEDIA AND POLITICS

Politics and government have always been shaped, in part, by the needs and aspirations of authoritarian forces. Today we take satisfaction in the public benefits from virtually universal access to mass media. But what would the future hold if, in the coming decades, media were to contribute no more than they do now? Anyone who has lived through the social upheavals in reaction to the Vietnam war or through the struggle for racial equality after the Supreme Court outlawed segregation as unconstitutional on May 17, 1954, must be concerned with that question.

It profits society little if more goods are sold every year, while an increasing number of black people lose faith in democracy. If the ghetto inexorably grows, physically or mentally, can our democracy survive? If the nation's children are more impressed with media glorification of antisocial behavior, what hope is there for the future? If the old or the poor are bypassed by the parade of inane escapist offerings of the media, are any of us secure in the faith that any segment of the population *is* immune from being so ignored sometime soon?

There is an obvious need for a new conceptual approach whereby goals are set for the media. Without limitation of the freedom to discuss or present anything acceptable to reasonable people, the media should be judged on *how well they systematically attempt to cover essential stories,* that is, on *how adequately subjects connected to poverty, children, health, education, religion, the environment, research, and so forth, are covered on a continuing basis in some systematic fashion.*

No less important is the consideration of *how well the media resist the temptation to glorify actions which, by any reasonable interpretation, are and have been considered undesirable and threatening to personal liberty and national cohesion.*

THE MEDIA AND VIOLENCE

If there is one policy worse than the censorship urged by those who regard what is odious or evil as nonexistent if hidden from public view, it is advocacy of the idea that nothing should be kept from public view. Yet some communicators contend that media presentations that are inherently antisocial or devoid of benefit to the community should be regulated only by the existing competition in the marketplace of ideas. Further, they argue that all publicity somehow liberalizes tastes and liberates the human race.

Often, champions of such uncontrolled communication are self-serving managers who deal in the manipulation of temptations which threaten the existence of the community. A result of their work is that liberty and license are made to appear synonymous and mutually dependent. Yet we know from countless historical experiences that there must be reasonable checks on that which is without redeeming qualities (murder as sport, and torture of the innocent, for example). As a general rule, reasonable people who believe in maximum democratic freedom do not support either irrational censorship or irrational dissemination of ideas.

The press of a free society is at liberty to display the defects of its system; the press of a controlled, authoritarian society is not. The subject of violence offers the best illustration of the difference. We know that dictatorships tend to thrive on general repression and specific uses of violence to control individuals and groups. Secret police, prison camps, and torture chambers are basic to the violence of dictators, who repress and hope to obliterate all manifestations of protest against state activities. Given their paranoia about constantly festering forces arrayed against them but hidden from sight, terrorism becomes a standard governmental tool. Yet despite its pervasive influence on the lives of the people, it is scarcely likely to receive coverage in the controlled press.

In a free society, protection of the people against most forms of violence is the first job of governmental organizations—executive, legislative, and judicial. The press is eager to make people understand the varied manifestations of violence, as in international wars and the instances of brutality we face every day.

In a democracy, the violence of the ghetto, where common conditions include ignorance, deprivation, and official governmental failure to ensure that the laws are executed in letter and spirit, is actually more overt than is the day-by-day violence in the average well-run authoritarian state. Violence is a seam-thread in the lives of migrant workers, of orphans in badly run institutions, of elderly people tied with sheets to wheelchairs in nursing homes, of soldiers mustered for senseless wars far away from home, of youth adrift in a technological world, and of students who protest what they have been taught to consider bad and who are outcasts for their actions and attitudes. Also, from industrially fouled air there is violence to the organs of the bodies, to the spirit, to ambition, and to community aspirations. These topics offer our media much material for exploration and explanation.

Robin Day, the highly experienced British television journalist, writes in his "Troubled Reflections . . ." that he knows the value of television coverage of violence in the world. He cites as examples the day-by-day depiction of the Vietnam war; the rages and the terror in Northern Ireland; the Biafran rebellion against the Nigerian central government, which brought military carnage and scenes of starving children; the peaceful protests of the blacks for equality, which led to new understandings only after the cruelty and brutality they encountered; and the invasion of Czechoslovakia by the Soviet Union, which laid waste Czech efforts to restore democratic government. All these stories of the 1960s, and countless others, featured violence in one or another of its manifestations.

Day is troubled because one mass medium, television, is particularly Janus-faced. "Television, which can encourage a pacifist revulsion against war, may also be an incubator of violent revolution." In respect to the civil rights struggle in the United States, he notes that "ideals are infectious and television has helped to spread them with explosive consequences." Television, he reminds us, was one of the key media used by the Czech government, before the Soviet invasion, to speed the momentum of the drive for independence and freedom. "TV helped create a situation which led in turn to a major act of international policy, the Russian invasion." For the inhabitants of Northern Ireland, television prompted changes of outlook about the world and about home-bred bigotries and alternatives. Television "helped stir a challenge to blinkered bigotry [and] gave impetus to the civil rights movement."

Day is concerned that television's vivid depiction of certain situations triggers unplanned, unforeseen, and sometimes unfortunate changes as well as those that are desirable. He wonders whether the most powerful mass medium tends to *rise* to the social and political occasion by encouraging education and informed debate or whether it tends to paint the great political movements and moments with broad strokes and heavy colors, in artistry which does not go beyond thematic depiction. He decides that "television is inherently incapable of giving fair and balanced reporting of a very large part of the world today. For every big television documentary which presents the facts behind a story, there is a myriad of bits and pieces on "fashions." "Easy publicity on television has helped the spread of what might be called today's 'fashions' in adolescent violence, whether by students or other minority protesters, skinheads or Saturday football vandals.'' [5]

Daniel Bell, a Harvard University sociologist, adds a most important consideration, one that lies behind most of the political *traps* of the 1960s when violence was written about or shown on television and movie screens:

> Violence was justified not only as therapy, but as a necessary accompaniment to social change. Watching the children of the French upper bourgeoisie mouth the phrases of violence and chant from Mao's Little Red Book in Jean-Luc Godard's *La Chinoise,* one realized that a corrupt romanticism was covering some dreadful drive to murder. Similarly in Godard's *Weekend,* where a real slaughter of real animals takes place, one realized that the roots of a sinister blood lust were being touched, not for catharsis but for kicks.[6]

ONE ROAD TO REFORM: A SYSTEMS ANALYSIS APPROACH

In response to the trash and the increasing quantities of ''filler'' materials of no social consequence which crowd television and radio schedules, people who advocate improvements in programming have generally neglected to offer suggestions for progressive change. That failing may be due, in part, to a preoccupation with specific content and avoidance of the crucial question: *What gauge or measure or formula should media managers use to try to make the media more responsive to public needs?*

To answer that question, one turns to advances made by scholarly practitioners who work in the area of national development analysis. David Easton, one of the more perceptive *systems analysts,* urges an examination of the physical, biological, social, and psychological *environments* of political life.[7] Suppose a television station owner or a network potentate decided that all programming, other than a certain portion devoted to the needs of the commercial buyers, would be divided into these categories and 25 percent of the available time would be apportioned to each of Easton's environmental areas. Such a plan would be certain to produce a radical change in programming, for without the systems approach, today's media managers can generally ascribe social or political importance to any content that by any stretch of the imagination is not clearly commercial.

The noticeable trend in the management of popular newsmagazines, such as *Time* and *Newsweek,* toward the establishment of somewhat constant categories or topics repeated weekly, reflects simplistic systems analysis concerns. Material in the October 15, 1974, issue of *Time,* for example, is grouped into the following categories: cover story, essay, art, books, cinema, economy and business, environment, modern living, music, people, press, religion, space, and theater. In addition, there are the categories of nation and world for coverage of events. Finally, there are letters from readers, and what are called "milestones."[8]

Although a similar approach has been taken by directors of individual programs, frankly offered as television "magazines," the medium more commonly appears to be a jigsaw puzzle whose pieces do not form a picture.

R. Vincent Farace, who calls mass communication a "kind of social lubricant," offers another systems analysis approach. He recommends tracking of *variables,* which are interdependent in that "change in one variable will produce ripples of change throughout the system." He finds it useful to set up an arbitrary categorization of nine developmental variables: mass communication, political system, health and nourishment, agricultural productivity, climate, population characteristics, cultural factors, economic factors, and sociopsychological factors.[9]

Suppose television managers decided to devote half of the total available time to socially significant programming and to assign staff personnel accordingly. In addition, assume that each of Farace's categories were given either equal time or fully adequate coverage. Add the requirement that the staff determine scheduling priorities for each category and documentation. Several results, given present conditions, might follow. First, there is the possibility that the attempt would break down virtually at the start because of inadequate resources. Second, there might be an immediate contest between the old guard, devoted to maximum profits with minimum efforts, and the advocates of systems analysis as the basis for scheduling. Also, cynical professionals might oppose the idea that mass communications is essentially a public service area. On the other hand, a genuine application of systems analysis techniques might cause truly radical changes in the public interest.

Current practice appears to be dependent on a series of benign accidents, with a smattering of structured ideas about news and public events coverage tacked on. That approach must be discarded if we are ever to do something about the danger to the public as perceived by Max F. Millikan: "The underlying causes of the recurring crises are rarely explored [and] the forces building up towards future crises are rarely exposed so that the public may be prepared for the next shock; no account is given of what might be done to avoid or alleviate these crises." [10]

IS VIOLENCE ENTERTAINING?

Physical violence, unchecked by what democratic people call the "rule of law," is the basic element in every form of despotism.[11] With little regard for this truth, the majority of television's managers conclude that exciting programs frequently depend upon the showing of brute strength. More and more, they see to it that such excitement depicts some manifestation of sadistic behavior. An ever-increasing amount of violence on television, excused privately by industry executives as necessary if they are to be competitive in the business of building audiences, is offered as entertainment. In the period from 1968 through the latter part of 1972, for example, "eight out of every ten programs contained some violence." [12]

Leo S. Singer, the president of the Miracle White Company of Chicago and a leading television advertiser, reacted to violence on television after becoming aware of its dangerous effects. On October 5, 1973, Singer, attempting to choose some worthwhile program to watch, turned to four different channels and found that he was faced with four shows with violent plots. Still seeking relaxation, he picked up his newspaper and read about Evelyn Wagler, a twenty-three-year-old social worker, newly arrived in Boston, whose car had run out of gas. She walked to a service station and was taking some gas in a can back to her car, when young hooligans pounced on her and forced her to douse herself with the fuel. Then they lighted a match to ignite the gas, and she burned to death. A similar incident had been televised three days earlier in a movie called *Fuzz*—a tramp was murdered in essentially the same way. Did the television film so impress some members of the audience that they actually took that terror to the streets of Boston? Singer thought so. Convinced that he had to do something corrective, he decided that his company would "no longer be part of that potential murder syndrome." He vowed: "As long as I was president of the Miracle White Company . . . we would never advertise on another violent show again." He was reacting against what he called the "oversaturation of violence," and he pointed up the need for "self-limitations on extremes." As he assessed the situation, "It is almost as if a TV college for the young, the mentally unbalanced or the criminally predisposed is being conducted on some of these shows. The criminal act is portrayed in such detail—and in most cases, carried out with such cleverness—that it is simply a job of monkey-see, monkey-do for the prospective criminal to react."

Singer attributed the major fault to circumstances such as those existing in his own company:

> Advertising was turned over to the marketing people and other experts in the field. Most of these people, as you know, purchase time on a cost per thousand basis. How many people can they reach with X number of dollars. Generally, this is done without regard to the type or content of the show. It is a rating game—win with profits. And to be perfectly honest, as long as our company's products were doing well, I never really gave it a second thought.[13]

Singer appeared before the U.S. Senate committee headed by Senator John O. Pastore, which investigated violence on television. A concern of the committee was the relationship between antisocial behavior and the welfare of people who are unable to *handle* the gross amount of violence portrayed on television. Bertram S. Brown, of the National Institute of Mental Health (NIMH), reported to the committee on research bearing upon television's impact on "cognitive and emotional development of children." The depiction of violence is not in itself the central problem, he observed. The problem concerns the amount of such programming and the cultural and situational effects. Quantity and quality are interrelated. Brown offered the view that "showing violence but not showing its aftermath, grief, misery, loss, what appears as a cause of violence is a critical issue."

These considerations appear even more critical as we pay proper attention to the conservative estimate that at least 1.4 million minors (under eighteen years of age) have severe psychological problems and 10 million more need some degree of psychiatric help.

NIMH, in cooperation with other groups, is sponsoring research on an "index of violence." Also in preparation is what is called a "multidimensional profile" to provide further precise measurements of violence on television. The Social Science Research Council, a national consortium of major professional social science organizations, is sponsoring research into qualities of violence in terms of "how vivid, how brutal, how illegal, how unjustified" it is.[14] Given the finding of the National Association for Better Radio and Television that a typical youngster between five and fifteen years of age saw the "violent destruction of more than 13,400 persons on television" in 1968,[15] we all need to be concerned.

George Gerbner, dean of the Annenberg School of Communications, University of Pennsylvania, who is a leader in the attempt to develop a violence profile, reported his findings to the Pastore committee: 78 percent of the leading male characters featured in network programming between 1969 and 1972 were involved in one or another violent activity, and 44 percent of the women characters were similarily engaged. He found that characters in television plays (including cartoons) were slightly more likely to be cast as victims if portraying nationals from other countries. Also "lower class characters, especially women, are most likely to be cast in the roles of victims,'' followed in frequency by upper-class and then by middle-class characters. ''Nonwhites, especially women, suffer more violence than they inflict'' in television plays.

Gerbner and his associates reported, in the summer of 1974, additional information connecting television characterizations to the perceived concepts of *reality* among people who are given to heavy viewing of television drama and comedy. His data are drawn largely from ''network dramatic programs transmitted in prime time (7 through 11 P.M.) and all day Saturday for one full week in the fall of each year after the launching of the new season.'' In six years of investigation he and his co-researchers analyzed ''656 programs (plays), 1907 leading characters, and 3505 acts or episodes.'' One objective was to find out what adults and children who were tested ''think is the true state of affairs.'' An important conclusion was that ''heavy viewers . . . carry their television experience into the real world of social reality when they give a 'television' rather than the very different 'real world' answer to questions on population density, employment, crime and violence, and law enforcement.'' [16]

MEDIA AND CIVIL DISORDER

How adequately do the mass media serve the public interest at times of civil disorder? We are just beginning to understand the essential problems and the meanings of recent events. In particular, we are trying to find out if the media can do more than they have done to reduce tensions at times of crisis. Do news presentations which stress timeliness tend to renew tension? There are significant clues before us now! [17]

The racially based summer protest riots that swept many American cities in the 1960s prompted theoretical analysis and practical reporting by the National Advisory Commission on Civil Disorders appointed on July 27, 1967, under the chairmanship of Otto Kerner, then governor of Illinois. By and large, the description of its members as ''moderates'' is justified. Thus, fiery radicals of the civil rights movement were excluded. This matter of membership is important because the Commission's conclusions are *temperate,* given the very bad situations it was examining.[18] Acting in a fact-finding capacity for the purpose of advising President Johnson on means of reducing tensions, the Kerner Commission arrived at these principal conclusions:

> 1. The nation is rapidly moving toward two increasingly separate Americas.
>
> Within two decades, this division could be so deep that it would be almost impossible to unite: a white society principally located in suburbs, in smaller central cities, and in the peripheral parts of large central cities; and a Negro society largely concentrated within large central cities.
>
> 2. In the long run, continuation and expansion of such a permanent division threatens us with two perils.
>
> The first is the danger of sustained violence in our cities.
>
> The second is the danger of a conclusive repudiation of the traditional American ideals of individual dignity, freedom and equality of opportunity. . . .
>
> 3. We cannot escape responsibility for choosing the future of our metropolitan areas and the human relations which develop within them. It is a responsibility so critical that even an unconscious choice to continue present policies has the gravest implications.[19]

These ominous views, given to President Johnson and thereby to the nation in February 1968, proved politically difficult for him. Very few concrete actions were taken by governors, media people, or the citizenry, despite general acceptance of the Commission's warnings.

President Johnson's original charge to the Commission included the question: What effect do the mass media have on the riots? Its responses to that and other questions serve students of the media well, even if President Johnson and his successors were less than eager to deal with the report.

The Commission undertook a vast research effort. Field survey teams interviewed "government officials, law enforcement agents, media personnel and ordinary citizens." Ghetto residents were specially queried. Regarding the specific period of rioting under study—the summer of 1967, with particular attention to a two-week period in July when the "worst came" in the cities of Newark and Detroit (with chain reactions set off—in neighboring areas) the Commission ordered a "quantitative analysis of the content of television programs and newspaper reporting in 15 riot cities" for the riot periods and the days immediately before and after. Also, a conference of media leaders from the newspaper, newsmagazine, and broadcasting industries was held from November 10 to 12, 1967.

The media got good marks for overall intent and coverage. Credit was given for the genuine attempt to provide factual and balanced reportage. But it was concluded that the total result was "an exaggeration of both mood and event," and the media were criticized for failing to deal with root causes and consequences of race relations problems.

A big problem, according to the Commission, was that reliance on media coverage of the disorders tended to produce the impression that they were worse than they actually were. However, the Commission members were pleased to find that sensationalism, overplay of violence, disproportionate time given "emotional events and militant leaders" were not *as significant* as they had believed: "Of 955 television sequences . . . 837 could be classified for predominant atmosphere as either 'emotional,' 'calm' or 'normal.' " Of these, 494 were classified as calm, 262 as emotional, and 81 as normal.

Newspapers came out well: "Of 3,779 newspaper articles analyzed, more focused on legislation which should be sought and planning which should be done to control ongoing riots and prevent future riots than on any other topic." Despite the newspapers' interest in legislation, however, there were serious flaws in their coverage. "Scare" headlines covering mild stories were cited. And it was found that some reporters often sought out factual material on the scale of the disorders from local officials who were generally inexperienced in such situations and tended to do a poor job of separating reality from rumor.

Given the fact that most of the trouble took place in all-Negro neighborhoods, television depictions of black-white confrontations or "race riots" were not precise. The sounds and sights transmitted on television (flashing lights of police cars, screaming sirens, burning buildings) caused what the Commission called "a whole sequence of association." Individual items of coverage were not necessarily inflammatory, but in association lurid imageries of whole situations could be triggered. The Commission also speculated on the effects of long years of media coverage of the civil rights struggles. Were media audiences "conditioned" by previous reportage? Were reactions "heightened" by free association with other events dimly or unclearly remembered? [20] On the whole, television, despite conscious efforts at objectivity, troubled the Commission:

> Television newscasts during the periods of actual disorder in 1967 tended to emphasize law enforcement activities, thereby overshadowing underlying grievances and tensions [and] tended to give the impression that riots were confrontations between Negroes and whites rather than responses by Negroes to underlying slum problems. . . .
>
> About one-third of all riot related sequences for network and local television appeared on the first day following the outbreak of rioting, regardless of the course of development of the riot itself.

Newspaper coverage showed some peculiarities. For example, when a riot was taking place in a newspaper's own city, the emphasis tended to be on the national rather than the local situation. Reports of troubles in other cities were featured; more headline attention was given to outside troubles than to the homegrown. "About 40 per cent of the riot or racial stories in each local newspaper during the period of rioting in that city came from the wire services." [21]

Television and newspapers were criticized for ignoring the fact that a substantial portion of their clientele is black. There is a good deal of bias in the constant display of what amounts to a white world. There is inadequate coverage of black people's activities. There is a failure "to portray the Negro as a matter of routine and in the context of the total society." [22]

To move beyond the riots of 1967 and deal with the larger problem, the Commission recommended new efforts along the lines of recognizing the lives and needs of minorities; integrating minority news and views into the total news effort; training black journalists for media work, especially with a start in urban affairs; and setting up placement programs for the newly trained.

Columbia University's School of Journalism developed a program in 1968 for the training of journalists from minority groups. Supported by the commercial television networks, the Ford Foundation, local stations, and newspapers, it was a positive response to the recommendations of the Kerner Commission. The program, at an estimated cost of $12,000 per student, provided for a stipend for living expenses and assured job placement upon graduation from the eleven-week course. From its inception to its demise in August 1974, 225 minority-group journalists were trained. Most were placed with television networks and newspapers across the United States.

Fred Friendly, former president of CBS News and now a key professor in the Columbia School of Journalism, reacted to adverse criticism of the last students in the program by labeling the program "discriminatory" because of "other minority group students who pay $5,000 to attend the regular graduate program but who do not get stipends or guaranteed job placement." What precipitated the end of the effort was the drying up of financial support from the media. As Robert Maynard of the *Washington Post,* and director of the program, said, "Now that there are no riots, we can't get any action." He declared, "If this program doesn't continue, then we go back to the old system in which minorities simply weren't hired and it will mean that a lot of brilliant people simply won't get in."

There was, of course, cause for legitimate criticism of the program. Some of the trainees were not college graduates, and this made them different from students in the rigorous regular graduate program. Moreover, students who had to pay $5000 for the regular program, which provided no guaranteed placement after they earned the highly respected master's degree in journalism, had some grounds for skepticism. The major criticism, however, should be made in behalf of the general society. The program was too limited, too short in time, and too much a psychological palliative for the media themselves.

The political lessons have still to be learned. Denial of opportunity because of current or previous conditions of servitude is fundamentally wrong. The cure will not be in the form of model programs set up to handle a lucky few. When the program ended, CBS had already announced its own minority training program, designed to produce five members of the communications fraternity in a two-year period.[23] Impetus for these too little and very late programs did not spring from carefully tailored efforts to match political and educational planning to the real needs of the deprived in this imperfect democracy. They must be considered makeshift reactions to the violence that poured from the ghetto streets onto television screens and into newsprint.[24] Ralph Nader warns:

> We need much more than just an episodic reform, an eruption of concern which gets little action and then spills over into dissipation and dispair—we need the kind of entrenched career role that begins carving out these interests in the crucible and decision systems—for good: a durable type of reform and mission that will not just replicate the undulations of reform throughout our history.[25]

CHILDREN AND TELEVISION

The mass media can be as important as the formal educational systems of our states, cities, and towns in preparing children for life. Indeed, they can do a better job when it comes to certain problems. For example, millions of preschool children are coming to rely on television for most of their important information about the world, and for most of their entertainment. Mothers and fathers, for better or worse, are propping up infants before television sets to keep them quiet and occupied. For those locked in poverty in the ghettos of the nation, television is the only way to see other parts of the world. Youngsters from rich and poor backgrounds watch television hours each day, no matter what is on the screen.

These facts do not appear to have impressed the majority of radio and television executives sufficiently to make them assume responsibility as educators or educational administrators. On the contrary, they are, on the whole, servants of the major commercial advertisers.

The protest against solicitations to or about children is a matter of the greatest social importance. Should we fail to secure corrective measures against present advertising practices, then we will most certainly fail to convert television from a primarily commercial medium to a primarily cultural one, in the broadest sense. After proper deference to advertisers' support of the media, one concludes that democracy is not uplifted by the overall tone and sweep of television advertising to children if the program content is twisted to meet the rules of moral aggrandizers in the profit-and-loss profession.

The bulk of such advertising aims at selling goods by stamping impressions on young minds. Our children are made to *see* what they are urged to buy. So much effort goes into stimulating the movement of commercial products that, by contrast, children find the most important problems of the day obscure and intangible. Youngsters who watch television are seldom taught that the world spins around more than the toys, cereals, clothing, and entertainments that are advertised constantly. It is obvious that very little effort has been made to present news programs for children, according to their needs, abilities, and interests. They must learn about news events from the programs prepared for adults by local stations and networks.

To be sure, it is a fundamental belief of most American parents that their offspring must be protected from fear, terror, evil in any shape or form. Unfortunately, there was little attempt to shield the children of Indochina. One manifestation of the news reports carried by television, in the whole of the Vietnam war period, was that American parents and children were spared the sight of almost all the horror visited on the people of Southeast Asia. Media managers took the part of the parents, protecting the vast audiences from the worst nightmares.

No case is made here for frightening children. Far from it! The innocence of children may be the only insurance against total disaster that we all have in a nuclear age. Even "statesmen" bent on war have children and grandchildren. Nevertheless, suffering cannot be shunted aside. Starvation cannot be pretended away by television schedules too crowded with fancies to make room for fact. Children must learn to *see*, to comprehend the substance of the grave crises in the world today. Otherwise, given what we know about human history and the present global situation, hope will be tossed aside in favor of candy bars.

Recent studies provide disturbing data relating to what children actually view by way of learning about the world. Professor F. Earle Barcus, of Boston University's School of Public Communication, testified before the Federal Trade Commission in 1971 on the intensity of advertising directed at children. His research was supported by the Action for Children's Television organization, headquartered in Newton, Massachusetts. That very active, socially motivated group has taken a special bead on Saturday morning children's programming on all three national commercial newtorks.

Barcus videotaped programs shown on four commercial television stations in Boston, affiliates of each of the major networks, and one independent UHF station. He worked on four Saturdays between May 29 and June 19, 1971, monitoring one station each week. In that survey he amassed 18¾ hours of tape for his content analysis studies. He found a heavy concentration of advertising (88 percent) for toys, candies/sweets, cereals, and foods/snacks in the six categories of commercial products touted. According to Barcus, "Commercial messages accounted for approximately 11 minutes per hour on the average." He calculated that if the 406 commercial messages he recorded were "spread out evenly over all time monitored, this would amount to one commercial message every 2.8 minutes."

Whatever the origin or format of the "Kidvid" Saturday programming (68 percent of the material "was from the network, 19 per cent recorded, and 13 per cent live; 88 per cent was classified as entertainment and 12 per cent as informational; 66 per cent used animating"), Barcus discovered that violence was a very important ingredient:

> Of approximately 10½ hours of *dramatic format* programming (stories, cartoons) 32 per cent was devoted to segments in which crime or its solution was the predominant subject matter, another 16 per cent involved interpersonal (usually physical) struggles between characters, and 16 per cent . . . the supernatural (dealing with ghosts, magic, witchcraft). . . . There were no stories about domestic (home and family) problems, religion, race or nationality, education, business and industry, or literature and the fine arts.[26]

Is this saturation with simulated violence harmful to young viewers? The recent report to the Surgeon General, U.S. Public Health Service, entitled *Television and Growing Up: The Impact of Televised Violence,* indicates a causal link between television presentations and aggressive behavior of youngsters. Although the report is said to have emerged as a compromise text which reflected a softened message agreeable to network interests, twelve very learned men and women with high research reputations concluded:

> First, violence depicted on television can immediately or shortly thereafter induce mimicking or copying by children. Second, under certain circumstances television violence can instigate an increase in aggressive acts. The accumulated evidence, however, does not warrant the conclusion that televised violence has a uniformly adverse effect nor the conclusion that it has an adverse effect on the majority of children. . . . The evidence does indicate that televised violence may lead to increased aggressive behavior in certain subgroups of children, who might constitute a small portion or a substantial proportion of the total population of young television viewers. . . . There is evidence that among young children (ages four to six) those most responsive to television violence are those who are highly aggressive to start with—who are prone to engage in spontaneous aggressive actions against their playmates and, in the case of boys, who display pleasure in viewing violence being inflicted upon others.
>
> The very young have difficulty comprehending the contextual setting in which violent acts are depicted and do not grasp the meaning of cues or labels concerning the make-believe character of violence episodes in fictional programs.[27]

One finding termed "most striking" was that research indicated "an increase in prosocial behavior among the children who viewed the prosocial programs (e.g., 'Misterogers Neighborhood'). This increase was limited to those young viewers who came from families of low socioeconomic status. These children tended to become more cooperative, helpful, and sharing in their daily relationships with others; the children from families of high socioeconomic status did not.'' [28]

Examples of Imaginative Children's Programming

Children's programs that get beyond the needs of advertisers to those of the young audiences have had only limited scheduling in recent years. There are, however, indications of a sense of social responsibility that deserves commendation.

''Misterogers Neighborhood,'' a National Educational Television staple for very young children, has won its creator and star Fred Rogers many awards, including the Peabody. Roger's low-key approach to children of tender age demonstrates his respect for, and appropriation of, his audience. Each individual program in the series is based on a subject such as ''why you keep clean'' or ''why people should respect each other.'' [29]

''Sesame Street,'' created by the Children's Television Workshop and first presented in 1969, is considered a real step forward in establishing a new genre of educational entertainment. The federal Department of Health, Education and Welfare, the Ford Foundation, and the Carnegie Corporation have contributed millions of dollars in support of the series. One of the publicized contributions made by ''Sesame Street'' productions is alleged progress in helping disadvantaged children of the nation read and write, or at least in helping them to begin the process of substantive learning at home.

The commercial networks have done some good work in the area of animated cartoons. In the 1972–1973 season the Columbia Broadcasting System presented ''Fat Albert and the Cosby Kids,'' which dealt with juvenile sociological situations in a way designed to appeal and make sense to children. The American Broadcasting Company produced ''Yogi's Gang,'' in which Yogi ''faces such enemies of nature and man as Mr. Pollution, the Envy Brothers and Charlie the Cheat.'' ''Children's Film Festival'' (CBS) and ''Take a Giant Step'' (NBC) are other examples of fine public service efforts.

Local television stations also have made progress on the quality side. KPLR-TV in St. Louis, for example, won the plaudits of parents, educators, and children with the introduction of ''Mindstretchers,'' a program for the seven- to fourteen-year age group.[30]

As interest develops in the creation of new children's programs, criticism is more pronounced on the work done. ''Sesame Street'' viewers, it is alleged by some, may not learn to a degree which would be applauded by professional educators. Some claims made for the series—for example, that it helps the disadvantaged—are rebutted. A black theatrical producer who lives in Harlem called it ''bland, plastic programming for white middle-class America.'' Another black critic, with high school teaching experience, opined that ''You can know all the ABCs in the world and still hate people and still misunderstand people. *Sesame Street* is the same thing you get in a typical all American white elementary school.''[31]

BEFORE IT IS TOO LATE

The unfortunate preoccupation with simulated violence which so marks television's social and political artlessness signals even more dangerous public manipulation. Misdirection of young viewers, by turning them away from the realities they must encounter, is a first stage of a planned turning of the whole society away from democracy.

Politically, there can be no democracy if illusion-mongering becomes the elemental base for public communication. Politically, there can be no democracy if the *chief instrument* for mass communication perpetuates ignorance about the world or activity fosters acceptance of brutality.

One could pretend that a balanced overview of all media is called for in this search for the truths about media politics. The fact is that the print media today do a reasonably *adequate* job of disseminating divergent viewpoints spanning the spectrum of ideas. Later we shall consider, in the context of definite situations, problems of print media in connection with the play of politics. Problems are indeed substantial; nevertheless, it is the electronic media, particularly television, which could debilitate the polity.

There is a tangible alternative to today's television structure, which is based on major control of prime-time programming by giant networks. That alternative, that hoped-for mechanism for progress beyond the limited fare of today to virtually unlimited programming possibilities, is cable television—the wired system. However, this great medium for allowing many voices to be heard is still in a novelty phase, and local needs and tastes are still only minimally served.

Despite daily promises by promoters of cable television, who claim to give priority to community needs, we are still worried about a network situation which caters first to the aspirin tablet requirements of audiences numbering in the tens and hundreds of millions. All the fine educational channels notwithstanding, it is hard to ignore the fact that television does not replicate the print media. Television presents the news, but without its subtle complexity; it entertains, but it curbs variety; it caters frequently to intellect but hardly ever to elites of the worlds of the arts, sciences, technology, or government. Through sloppy treatment or lack of attention, television minimizes certain grave problems (such as the world population explosion) ad absurdum, while other equally potent problems (such as threats to democracy posed by a frequent playing with violence in programs that entertain) are made invisible to the citizenry through the constant use of cheap theatrical tricks.

One of the truly wise men of this century, the British novelist, scientific observer, technocrat, and political savant C. P. Snow, warned that we live in a "state of siege." Our world is closing in on us. How can it be, he asks, that we have begun to learn about DNA (the genetic code) through exciting research in molecular biology; begun to voyage about and on the moon and to probe the celestial regions; begun to tangle with the "universe of particle physics," and find the total effort of understanding mankind becoming more and more difficult? How is it, he wonders, that we see so much and are concerned with so little? Snow singles out television, to make a point about our knowledge of human suffering. "We know it is happening. We see people starving before they have died: we know that they are going to die. We see the evening's killings in Mexico City and same evening in London. We see the victims of famine in Biafra. We know it all. We know so much, and we can do so little. We turn away."

First, he reasons, the sheer complexity of the world frustrates most individuals. This and other causes of intellectual distress led Snow to the conclusion that we are all becoming more "callous about human life." Second, the callousness appears to increase as we all become aware that "in many places and for many purposes, there are already too many people in the world."

If there is any hope, Snow sees it in dealing with our options wisely, not ignoring human selfishness but not devaluing human intelligence. "We have to tell the facts. We have to make sure that people understand those ominous curves—the curve of population, the curve of food supply. We have to tell what the collision means." [32]

If Snow is not one of the supreme optimists in placing his ultimate trust in intelligence, then television—*the only true mass medium*—cannot be allowed to fondle us all into the abyss. Can one vent enough scholarly (if nervous) anger on this subject? While we play in our technological outpost called the United States, the world may be changing dramatically enough to make us alter our mass media to reflect reality.

We need more than documentaries on television: we need to treat each major problem as though it were a full-fledged curriculum, with audiences enrolled on a pass-or-fail basis. Consider again, for example, the subject of violence, and how rationality and brutality will collide thirty-five or fewer years from now, when the world population doubles. (In some poorer countries of this globe the doubling will take place in twenty years or less.) The founding fathers of the United States arrived at their sound conclusions about human freedom when this continent was overendowed with nature's bounty in relation to the sparse population. Now the worldwide population increase is estimated at 2 percent a year.[33] What will it be later? Resources of the earth are strained even now, and the great technological nations are well aware of their own disproportionate consumption of the total available.

Think of the consequences, in terms of potential violence, if the great populations of the developing countries continue to receive the lesser share while the relatively small populations of technologically advanced countries consume the greater. The possibility of a life-and-death confrontation between the haves and the have-nots provides ample material for but one full-fledged curriculum in which we are all, of necessity, enrolled on a pass-or-fail basis. Can television, *the* great instrument, continue to reduce this and practically all other significant subject matter to mere entertainment?

RATIONAL TELEVISION DOCUMENTARIES: EXCEPTIONS THAT PROVE THE RULE

Given the great need for helping the masses of national and international audiences to understand significant issues of the day, one would assume that television documentaries would be a major product of the industry. For reasons discussed earlier, this is not the case. Yet the disappointing volume of documentaries does not reduce the significance of the fine productions of recent years. Indeed, with such emphasis on commercialism in the industry, it is a wonder that television has achieved as much as it has in the area of documentaries.

The medium is only nominally regulated when it comes to insistance on quality programming. Radio and television share constitutional protections with the other media but are subject to controls imposed by Congress through its creation, the Federal Communications Commission. Under existing regulations, however, the FCC "does not prescribe the time to be devoted to news, education, religion, music, public issues, or other subjects. Programming can vary with community needs at the discretion of the station licensees." [34] The FCC is not now, and never has been, a force—moral or legal—for initiating quality public affairs work.

With no tradition apart from journalism to fall back on, and with a medium presently unable to compete in any real way with the print press, it is not surprising that television managers generally do as little as possible with costly in-depth public affairs productions. Nevertheless, the traditions of journalism do apply, principally because the first television reporters came primarily from newspapers. Moreover, the standards of quality reportage are constants, whatever the medium of communication may be.

Documentaries are among the better features of television, showing evidence of systematic analysis of topics not otherwise attempted. Disappointingly, they are not numerous in any given year. For example, in the category of prime-time special programs, in the period from September 1969 to April 1970 only eighteen news documentary shows were presented. Audiences have not exactly cultivated a taste for the more introspective of these programs: when the Neilsen audience research organization rated the American Broadcasting effort "A Matter of Conscience: Ethics in Government," it was the least favored of 176 programs evaluated for audience appeal.[35] Other programs in the period were devoted to aspects of the National Space Administration's Apollo flights; indeed, of the eighteen news documentaries, eight dealt with Apollo. An innovation was the combining of material previously broadcast on network newscasts, to form a hard-biting look at an American fighting unit in Vietnam: "Charlie Company," based on John Lawrence's reporting for CBS, contributed much to public understanding.

Two magazine-format series, "60 Minutes" (NBC) and "First Tuesday" (CBS), managed to deal with some difficult matters and more glossy subjects. "CBS Reports" alternated with "First Tuesday" and added to the reportage on *hard news* matters. Controversial material was not avoided. Fearless of Pentagon disapproval, "60 Minutes" managed a good, scrutinizing report on problems connected with the Navy's cost-overrun development of the M-48 submarine torpedo. "First Tuesday" provided very professional reporting on the sad situation of American Indians and on the "nuclear establishment." Other documentaries were developed for, and consigned to, audiences in the nonprime-time hours of the week, such as the "Sunday afternoon 'ghetto'."

Among more recent developments, CBS in the spring of 1974

announced a new series of six news special on subjects of broad national interest. "The Food Crisis: Feast and Famine," first in that series, compared domestic problems of inflated food prices with increased mass starvation in the rest of the world. Well researched, and documented visually in salient manner, it was a credit to the professionals in the public affairs department of the network.[36]

Another CBS documentary, "The Palestinians," in June 1974 provided the audience with a good review and a carefully objective assessment of one aspect of the Middle East troubles.[37]

Social conditions faced by individuals trying to work out their problems was the basis of "Three Women Alone," presented by WOR-TV in New York City. The experiences of a divorcée, a widow, and a "liberated" young woman were examined as they tried to live full lives without males to assume a share of responsibilities.[38]

In 1974 actor Hal Holbrook was featured in a portrayal of Commander Lloyd M. Bucher, the captain of the U.S.S. *Pueblo,* the surveillance ship seized by North Korea in 1968. The magnificent interpretation, using a good deal of factual material, illuminated a very troublesome episode in recent history.

Of unusual interest here, because it involves a drama, is "The Autobiography of Miss Jane Pittman." The outstanding young actress Cicely Tyson played a fictionalized character who had lived through the elemental triumphs and tragedies of American black people in the course of her 110 years. She saw the pains of slavery and of the civil rights movement of recent years. The program was of great value because it brought to life what was otherwise lost to documentarians, who usually rely on available artifacts and traditional sources of historical proof. The sagas of black people were not summarized in "Miss Jane Pittman," but it came close to genuine historical significance as an interpretation of the travails of many black people. In short, this drama became a social document filling the gaps where history had to be revealed. A fictional interpretation, it surprised documentary purists by coming very close to truths about the human condition.[39]

Interpretations do not always gain general acceptance. In 1968, for example, thirty-three members of the U.S. House of Representatives protested the screening of National Educational Television's "North Vietnam: A Personal Report," made the year before by the British journalist Felix Greene. He was sympathetic to the North Vietnamese at a time when the American public was psychologically and physically mired deep in war. Following the filmed report was a discussion between David Schoenbrun, a distinguished foreign correspondent, and the equally noted Professor Robert A. Scalapino, of the University of California at Berkeley. Schoenbrun and Scalapino argued the then torrid issues, reflecting their personal concerns. The correspondent doubted the wisdom of our nation's Asian policies; the professor was sympathetic with the government's strategies. On the whole, difference of opinion educated the audience on essential questions.

Felix Greene was singled out by Dr. Walter H. Judd, a former congressman of passionately anticommunist persuasion, as a "propagandist for Communist China and Communist North Vietnam." Judd wrote to the President of NET, Mr. John White:

> It is highly improper for NET to permit its facilities to be used by those who distort the truth for the sake of promoting the objectives of anti-democratic governments that are the avowed enemies of the United States. If this is NET's concept of serving the public interest, then the public has a right to ask that the NET management be turned over to men who understand that the public interest is not served by the dissemination of half-truths and lies to confuse the American people for the benefit of our enemies.[40]

Considerable agitation was a by-product of another highly political documentary, "The Selling of the Pentagon," which was produced by CBS and telecast in February 1971. It was an investigative report on the public relations techniques and programs of the Pentagon. According to Richard S. Salant, president of CBS News, the subject warranted examination since it involved issues of "civilian-military relationships, the fine line between information and propaganda, and even war and peace."

The program offered few surprises to avid readers of better newspapers and magazines which had already covered the military's public contact methods. Thanks to television, however, what had been known to a limited few became common knowledge among the public at large. The Pentagon was far from pleased with the program's content, which raised doubts about the propriety of some solicitations by citizens' groups played up to by the Pentagon. The chairman of the House of Representatives Armed Services Committee suggested that the presentation was un-American. Salant noted that "the Vice-President of the United States [Agnew] . . . found us . . . disreputable."

In the midst of this controversy, in April 1971 the Investigations Subcommittee of the House of Representatives Commerce Committee issued a subpoena for the film and a transcript of what was broadcast, in addition to all the raw material not used by CBS in the final version. CBS resisted the request for the nonbroadcasted materials (untelevised film, notes, memos, contractual descriptions, records on the disbursement of money, and so forth) but agreed to provide the actual broadcast and text materials. There were other rather speedy developments. The National Academy of Television Arts and Sciences awarded the producers an Emmy in May, showing its high esteem for the program. The Commerce Committee voted to ask for the full House to hold CBS and its president, Frank Stanton, in contempt. More sense reigned on the floor of the House of Representatives, however, when the members voted by a substantial margin (226 to 181) to "recommit." By sending the contempt citation back to committee, they averted a clash between a free and responsible press and government officials on the wrong track.[41] It was a good sign of the times that so many congressmen were willing to give support to an overt hunting expedition mounted by officials who felt abused by the press in general and, in this case, by CBS in particular.

As noted, these documentaries are not the answer to the big problems raised about the media and civic education, but they are just about the best efforts of television today. If they are curbed, the electronic giant will truly fall asleep. What, then, becomes of civil rights? Who sounds the alarms to alert a free people to great danger? If the Pentagon cannot be criticized for its public relations, what will be the limits of inquiry into the common defense situation by the media and by private citizens?

Every really worthwhile documentary on any subject dealing with justice or power offends some individual or group. The "Selling of the Pentagon" affair merely shows the friction between giant organizations, each struggling to maintain its alleged rights. Suppose CBS had lost the battle? Considering media managers' increasing sensitivity to criticism, would it not logically follow that any type of investigative reporting could be condemned and halted by the exercise of intimidation?

Are we to be deprived in future of programs such as "Suffer the Little Children," of the 1971–1972 season, which showed how youngsters in embattled Belfast, Northern Ireland, were choking in the bitterness, hatred, and futility infecting their homes, streets, and country? Are we to be denied daring and necessary programs such as "The Blue Collar Trap" (NBC News), which displayed the disastrous private lives of certain young workers in auto assembly plants? Will efforts such as "Who Has Lived and Not Seen Death?" (WNBC), about dying people confronting their own death, or "Heroes and Heroin" (ABC), about drug abuse by soldiers in the war zone, be abandoned by frightened television reporters, producers, and managers? [42]

Richard S. Salant was absolutely right when he said, in his examination of the "Selling of the Pentagon" fight, that "tyranny does not always come by coups. Repression and erosion of the First Amendment normally roll in, like the fog, silently on little cat feet." [43]

NATIONAL DEVELOPMENT AND MEDIA GOALS

A definite dichotomy is clearly seen in any general inspection of the mutuality of the media and politics, and it is particularly glaring in regard to the great Pied Piper, television. Enough programs of socially redeeming value are presented to enable critics to show how important goals of the democracy are enhanced. Mirror opposites of those presentations are the cheap and sordid programs featuring views of the breakdown of the state, through the glorification of crime in all its aspects. Honest and well-intended critics are perplexed. When they eulogize "Zoom," the public broadcasting treat for youngsters, should they coincidentally launch a tirade against the mass of ''Kidvid'' trash stressing violence? Should a ''Masterpiece Theatre'' presentation of an installment of Galsworthy's ''Forsyte Saga'' be praised without an accompanying condemnation of the latest mayhem so vividly depicted on ''Hawaii Five-O'', the weekly series which plays out what purports to be a ''cops and robbers'' theme around stark cutthroat incidents. Lawyers like to speak in terms of the *weight* of evidentiary material. On that basis, the weight of evidence forces the conclusion that we run the risk of destroying our democracy if we continue to pretend that the socially reprehensible is balanced by the socially laudable. Perhaps it is true that we are training children more for acceptance of cruelty in daily life than for its rejection. Perhaps we are permitting mass marketing of ''entertainment'' violence on such a scale, by and on behalf of commercial interests, that we no longer have the option to turn it off without a very hard fight. Perhaps our children are already so conditioned for violence as a routine that they, as adults, may not understand the critical distinctions between dictatorships and democracies.

Is there a way out? Can *intervention* through planned information, discussed earlier, really work for the democracy?

One of the curious facts of recent decades is that American developmental experts, advising what are called ''less developed countries'' (LDCs) in foreign-service jargon, have had no hesitancy in advocating media, especially television, as the natural benefactors of those interested in intranational democratic social construction.

We set the stage for this discussion by posing several relevant questions:

1. Can we relate media research on LDC problems to media problems affecting large groups at home who are politically deprived?

2. Do media research findings about problems overseas help us understand the new directions for media reform in this country?

3. What objectives should we press for first in the campaign for better media and better politics? If information dissemination ramifications

on the polity are so influenced by the angles of societal *thrust* set by media managers, can we change these angles in order to intervene more directly into certain problem areas and thereby encourage new social and political outcomes?

For more than two decades, television as a developmental tool was accepted by United States communications experts as one of the basic keys unlocking socially progressive change in any society. Looking back on their work, especially on their enthusiasm in the 1950s and 1960s, one wonders if a way out of our present dilemma is to be found in their developmental approaches. Be warned by this observer that the practical chances for such a developmental overview, at this late stage, are not likely unless the whole of the foundation for the present support of the television and radio media is challenged. This opinion will, no doubt, make many of the commercial overlords overtly oppose the whole of this text. They oppose any basic reform, and they are well defended in their views by their accounting chiefs, who know how big the bonanza is and how it can grow with each corporate reporting.

This introduction of the developmental thesis, based on the approaches taken by our foreign assistance communications advisers, is elementally dangerous to the exponents of what can be labeled the corporate entrepreneur thesis based on overwhelmingly commercial objectives.

One extremely lively and persistent individual who challenges commercial dominance of television is former Federal Communications Commissioner Nicholas Johnson. In his book *How to Talk Back to Your Television Set* he observed:

> Government is awakening to the dimensions of the task of identifying and enforcing the public interest as the communications revolution gathers force. But carrying out that task requires—especially in the design stage—a brand of imagination that does not flourish in bureaucratic offices, where weekly deadlines leave little time for reflection. The challenge is to make technical advance serve human needs, to define those ends, and mold the techniques accordingly.

He made those remarks after commenting on the possible future of cable television in the United States.[44]

For Wilbur L. Schramm, the doyen among advisers to LDCs, any communication should be in the context of a "planned dynamic; messages should stimulate or reenforce "a felt need." Practical benefits and the encouragement of cooperation are vital. Communications developments must be in keeping with economic and political decisions that fit the culture. Schramm asks:

> How fast do we want to go? . . . What ideology do we want to develop into? . . . What is the best mix, for a given purpose and a given time of mass communication, with interpersonal communication? . . . What is the best mix of campaigns and campaign goals? . . . What is the best distribution of resources so that communication can be made to flow where it is most needed? [45]

Schramm and his associates, great believers in television as a tremendous instructional force, have drawn up a number of specific plans based upon field research for agencies of the United States and for UNESCO. He is directly informed of the efforts to raise whole populations educationally in such diverse places as American Samoa, Peru, India, and Italy—and in Hagerstown, Maryland, where experiments with instructional television began in the local public schools in 1956.

Schramm believes that the media must be supported by interpersonal communication. He feels strongly that they need to be considered prompters: "It is not productive to think of the media as pouring content into viewers and listeners; a better way is to think of them as *stimulating learning activity*." [46] This stimulation can easily get out of hand, however, with profoundly dangerous political side effects, if television depictions are not part of a planned dynamic. Experts have observed a rise in aggressive feelings and behavior as a result of presentations which stimulate expectations for which there is little or no practical support. Poor people learn, for example, how badly off they are in relation to what they come to see of the world on television. In most countries they can but protest their condition without hope of redress of their newly articulated grievances.

Scholars also warn that there can be precious little political reform from projects which dote on social differences, either deliberately or by virtue of poor planning. The identification of other national peoples with traditional American ''images'' may be so weak as to be counterproductive to efforts to interest them in democracy via dependence on mass communications. Stress on affluence, says Harry Schwarz, disturbs audiences more attuned to messages emphasizing limited outside assistance and much self-help. Most audiences are confused by American insistance on the virtues of social mobility and social freedom. What comes through, all too easily, is a desire to obtain the material things seen on the television screens.[47]

If we substitute *community building* for the term *nation building,* so often used by media advocates, words of caution relative to autocratic leaders' interest in such work are in order. To destroy ghettos and improve the lives of their present inhabitants, powerful political and economic leaders must be willing to help the electorate change, uncertain how the better opportunities will affect their own situations. Autocratic elites do not look upon the mass media as ''potentially valuable vertical linkages between the rulers and the ruled.'' They fear the potentially disruptive effects of the media on their own monopolistic power.[48] In the United States, the suspicions of the press broadcast widely by some Nixon administrative people centered on alleged *abuses* by the press, the chief of which seems to have been its ability to make Americans question their governors.

It is notable that any developing society is at least a dual society, be it divided between the wealthy and the poor, the educated and the illiterate, technologists and ordinary laborers, inner-city residents and suburban populations. In any division, the media have the important political role of stimulating people to take action in conformity with new values. If that task is done poorly, there is no understanding between the different groups in society, and there is the strong likelihood that divisions will become increasingly fixed or frozen. Since the essence of democracy involves the political drive to better the condition of human beings, failure to communicate new values through the mass media indicates a crucial failure of doctrine and of system of government. According to Ithiel de Sola Pool, media can bring the modern world to the most remote "squatter settlements" as well as to the city dwellers. "Like all revolutionary forces, they prepare men's minds for new desires more rapidly than these desires can be satisfied." The media, if they are to "do more than add to frustrations, must be part of a program of sustained development." [49] In short, if a community receives new social values via the mass media, the new values must be simultaneously communicated on a broad front—in schools; through social work priorities; through provision for job opportunities and job security; through programs designed to help the weak, the old, the untrained; through recreational and entertainment programs. Otherwise, the new social values will strengthen old frustrations when all groups see them as supremely theoretical.

Even when the approach in communications is well coordinated with a major community force, such as the educational system, there are obvious pitfalls. One of the most significant efforts was directed in American Samoa in the 1960s. Out of a total population of approximately 26,000 (1968 estimate), about 7000 primary and secondary pupils were involved in a project designed to raise educational achievement by means of all the television teaching techniques then available. Support of the teachers was not neglected in this far-ranging program.

In this isolated environment in the Pacific, with a clientele under an educational television umbrella, television seemingly could produce nearly optimum results with the six channels operating by the autumn of 1965. Soon pupils were speaking better English, important content was being taught in a "more uniform way throughout the system," and educational opportunities were being "equalized." On the whole, mainland American educational goals were made the norm in Samoa, and the project achieved the intended results.

Of course there are political questions that never were fully resolved. The most important one was: Were we right in so intruding into Samoan culture? In the exercise of professional communications expertise, should we pay more attention to the society we want to uplift? It is not clear at this time—regarding American Samoa, American Indians, the black poor, or the white poor—how much of value is sacrificed in the name of progress. Cultural integrity, once successfully breached, will not be put whole again by communications. As was asked before, what ideological and political objectives act as the controls over the messages?

In this particular situation in Samoa, main sources of resistance appeared. They included United States educators who rejected the idea that television should "carry the core of the teaching," and spoke with considerable authority for some parents and teachers, as they held important positions in the Samoan system; "groups and individuals who felt their positions might be threatened by a large increase in better-educated, English-speaking graduates of the schools"; "groups and individuals who were concerned about changes in Samoan culture by an educational system built around television"; and teachers made insecure by the new methods and higher standards.[50]

Similar resistance was noted in evaluations of ambitious television programs sponsored by the United States in Colombia, Jamaica, and El Salvador, for example.[51]

On the basis of lessons learned from experiences in Brazil, the Ivory Coast, Japan, and Italy, the United States interagency task force working on telecommunications in the LDCs concluded:

> Television must be used . . . for achieving fundamental solutions to priority education problems. It must be thoroughly integrated to the educational system. . . . Planning must proceed from an analysis of the problems in local conditions, not from the blueprint for introducing a technology. . . . The systems characteristics of the educational process, with all of their interrelatedness, must be fully recognized in planning the use of television. Many technical, organizational and human elements must be given equal attention; for significant failure in any one of them will disrupt the entire system.[52]

Is there some value in looking at the United States as if it were a "less developed country" and planning its television accordingly? Industry leaders who insist that there is nothing drastically wrong would probably say no. After what we have been through in the last decade and more, what responsible citizen would not want major reforms in the interest of this polity? Even though present chances are slim that the developmental approach will be actively considered, it is an alternative. How much more must be endured before progress is made? As we wait, we must take hope in the old adage that "nothing can stop an idea whose time has come."

LEARNING FROM OTHERS

The author has been able to test one aspect of the mass media developmental thesis presented in this chapter. In December 1974 and January 1975, and in July and August 1975, he was consultant on media and national development, specifically asked to help revise the curriculum of the School of Mass Communications of the Malaysian government's Mara Institute of Technology (MIT). That college-level center started as a pilot program in 1964 with fifty students graduated from four professional courses of study. It has grown to a major educational institution which by 1974 was graduating more than five hundred students annually from twelve professional schools.

More than 80 percent of the students receive full government benefits (tuition, lodging, board, fees, travel allowance, textbooks, medical coverage) for the three-year courses of study. The Malay student body is drawn primarily from the rural villages of the country. The government's intention is to bring Malays into the middle-management professional life of the country in a real sense and thus to make the two major ethnic components of Malaysia—the Malay and the Chinese—comparable in twentieth-century technological terms. Although this special treatment accorded Malays has its political base in a Malay-dominated political structure, its benefits should not be denegated.

The Chinese live primarily in urban areas of the nation, and consequently tend to dominate commerce and industry. The Malays are primarily from the rural areas and must be brought ahead in all aspects of technology, commerce, and the professions. Accordingly, the government considers the first decade of Mara's history a phase in the development of the Malaysian story. After the Malays have taken their places in modern undertakings, it is hoped that no special treatment will be necessary and that a completely open enrollment will follow.

The first director of the Mara Institute of Technology, Dr. Arshad Ayub (at this writing, deputy governor of the National Bank of Malaysia), is a farsighted educator and developmental planner. Aided by energetic and visionary people such as the head of the School of Mass Communications, Puan Marina Samad, he has initiated progressive social change. Both Ayub and Samad are products of Malaysian and Western education, and they know firsthand the new procedures and techniques that have come from American colleges and universities. Professor Samad is a practical and theoretically sound young woman who has earned two master's degrees at American universities. Dr. Ayub was recently honored by Ohio University with a doctorate for his educational achievements.

With appreciation and strong support for recent desegregation efforts, we may still ask, What would happen if we in the United States faced the situation squarely and offered special advantages to the groups that have been unable to compete well enough, in order to alleviate their social, economic, and political distress? Suppose educationally deprived Americans—black, white, Indian, Mexican-American, Puerto Rican, and Cuban—were offered schooling equivalent to the Mara program, with the federal government footing the bills for sizable numbers of candidates considered sound prospects for middle-management technological and professional positions. Does this separatism run counter to *Brown v. Topeka* (1954), the classic U.S. Supreme Court decision on necessary racial equality? Obviously not, since we would be more adventurous than the Malaysian government and work on behalf of the many individuals from *all* constituencies who are unable to deal with the demands of the technological age. It would be realistic because the ideal of meeting real needs imposed by the academic stepladder would be transposed into practical programs.

In Malaysia, the effort to urbanize and train the rural Malays is not a complete success. Despite careful selection of trainees, a number drop out, unable to make the rapid personal adjustments of life style required or to meet the academic demands. In the United States as well, failures would have to be anticipated along with achievements.

In Malaysia, there are pessimists who deride the special training approach, preferring to concentrate on the current differences agitating Malays, Chinese, and Indians. Here in the United States there are social pessimists who prefer separatism among ethnic groups or who concentrate on how badly schooling changes have affected racial relations within the human race.

Students from MIT's School of Mass Communications are working daily, in on-the-job internships after academic classes, for all the major communications organizations of the country. Graduates will soon be taking their professional places in middle-management professional work. Very promising students have been selected for further training in the United States and other countries. The author was able to affirm personally that social exchanges are improving at the major communications enterprises as the trainees join their compatriots. It is only a beginning, but a very significant one.

Far away in Kuala Lumpur, officials worry about social tensions that can add to traditional factionalism and about the divisiveness that can tear society apart. Officials do the same sort of worrying in Washington, D.C., and in the state capitals and every city hall of the United States.

The special program in the School of Mass Communications at MIT, which is already important in basic political terms, is based on proven achievements in other schools of the institute, such as Accountancy, Art and Design, Architecture, Business and Management, Engineering and Computer Science, and Mathematics. It was begun in July 1972 with a mere forty students. Current (1976) enrollment of approximately two hundred students indicates the emphasis put on the effort.

It appears that this innovative program should be better known to American politicians, communicators, and educators.*

---

* In addition to those mentioned in the text, I should like to single out three people who actively facilitated my studies in Malaysia. They are Mr. Ahmad Nordin (former Director-General of Information Services Malaysia), Mrs. Shahareen Kamaluddin (of the faculty of the School of Mass Communications, MIT), and Mr. Haynes Mahoney (Public Affairs Officer of the USIA, Malaysia).

By Philippe Weisbecker. © 1973 by The New York Times Company. Reprinted by permission.

THE STRUGGLE FOR MEDIA FREEDOM

3

If freedom of the press signifies that governmental controls are minimal while guarantees of the rights and privileges of inquiry are maximized by law and custom, then the American people are blessed. If such freedom implies truly adequate and responsible media coverage of events and processes, as well as publicly beneficial defenses of the media by governors and politicians, then the American people have cause to worry.

Since the end of World War II, one reason after another has been offered by the officials of the increasingly administrative state for suppressing or controlling evidence or reports. *National security* has often been used as the convenient excuse to force controls on the press. We became accustomed to hearing about the presumed benefits of not knowing about significant atomic power developments, for one example. During the past decade and more, the Vietnam war has served as the greatest single excuse for restricting information. By the second term of Richard Nixon's Presidency, no citizen reading a newspaper or listening to the radio or watching television could be sure that the national news was complete or undoctored. "Executive privilege" was the popular slogan of the man in the oval office and of his subordinates throughout the executive branch of government. Newspeople and politicians alike, frustrated at the closing of sources of information, were increasingly united in the defense of democratic political orthodoxy, which requires the stimulation of ideas.

Justice Louis D. Brandeis said it well in 1927:

> Those who won our independence believed that the final end of the State was to make men free to develop their faculties; and that in its government the deliberative forces should prevail over the arbitrary. . . . They believed that freedom to think as you will and to speak as you think are means indispensable to the discovery and spread of political truth; that without free speech and assembly, discussion would be futile; that with them, discussion affords ordinarily adequate protection against the dissemination of noxious doctrine; that the greatest danger to freedom is an inert people; that public discussion is a political duty; and that this should be a fundamental principle of the American government.[1]

That Brandeis testimonial to the merits of public deliberation sounds strikingly familiar to many others on the same theme. When we see it as an extension of his and Samuel D. Warren's arguments in their now classic interpretation on "The Right to Privacy" (1890), a chord of logic relative to elemental current problems begins to emerge. The authors contended that the new forces of publicity had begun to interfere with privacy in a harmful way:

> The intensity and complexity of life attendant upon advancing civilization have rendered necessary some retreat from the world, and man, under the refining influence of culture, has become more sensitive to publicity, so that solitude and privacy have become more essential to the individual; but modern enterprise and invention have, through invasions upon his privacy, subjected him to mental pain and distress far greater than could be inflicted by mere bodily injury.[2]

Today, publicity seems even more of a threat, at least insofar as most citizens are concerned. They are less able to defend themselves against the ceaseless din created by the media than they were in 1890 or 1920 or 1960. This deluge of information often does not even include the basics one needs. Were Warren and Brandeis surveying society now, they would probably be warning us all to beware of the curtains of privacy drawn around high government officials in an attempt to establish enclaves against the "deliberative forces" and thereby conceal arbitrary action from media scrutiny. A perplexing problem has emerged. While the "man in the street" has surrendered more of his privacy to his government each generation, hoping that this intervention would be tempered in his favor by fair politics and a vigilant press ready to sound the alarms, power has become more important than public accounting to certain of his leaders. In particular, service in the executive branch of the national government seems to have reshaped some key elected leaders into supremely autocratic administrators. Such leaders, with little real humility, use the press to convey essentially self-serving statements about the so-called awesome responsibilities they shoulder for the rest of us.

Their plaintive appeals for understanding are not entirely groundless; indeed, those who hold the jobs of President, Secretary of State, Secretary of Defense, Secretary of the Treasury, and so forth, are not to be envied in this very troubled world. Nevertheless, there is nothing in common sense, law, or democratic tradition to permit officials wide freedom of action outside the glare of publicity.

That one phrase, *national security,* has been used so often to conceal official plans, policies, and actions that we no longer know its meaning. With so much evidence of inanity and corruption heavily censored before the public was informed by the press, no citizen can be certain that national security is in his best interest.

Sometimes, as we know, claims of privilege against publicity are made for convenience. It is an accepted fact of administrative life that keeping the public informed is a very demanding activity. Efforts directed toward providing explanations, arguments, and answers are extremely time-consuming and can divert officials from other important work. When a subject is highly controversial, the processes of democracy are a strain for even the best officials who are required to defend their procedures and policies, and possibly to redraft or reject their own ideas, after the public is informed of the issues.

Whatever sympathy we owe our leaders, they must no longer be allowed to avoid the glare of publicity. James Reston, one of the most experienced and respected of journalists and doyen of the Washington-based press corps, advised college freshmen in 1965: "The problem of this age is not so much insecurity as it is ambiguity. We are living in a time when change is the order of the day, when nobody but a fraud or a fool would pretend he had any perfect solutions, and when, above all, we need great integrity and flexibility of mind, not only to understand but to endure all this complexity." [3]

HUTCHINS AND THE PRESS

The erosions of press freedom since World War II have been significant enough to make the most responsible critics fearful that the days of representative government may be numbered. If, in a future national crisis, an unfortunate public discovered that weak or antidemocratic officials held the key posts in government, a quick slide into complete dictatorial control might not be averted by press criticism. Bear in mind that one of the essential political jobs of the democratic media of mass communication is to prevent the unsuited and the dishonest from coming to power, and, if that is not possible, to act in partnership with politicians, civil servants, the courts, and the public to control inadequate public servants.

Addressing the American Society of Newspaper Editors in 1955, Dr. Robert M. Hutchins, who had directed the 1947 work of the Commission on the Freedom of the Press, reminded his audience that he had spoken to the members a quarter of a century earlier. At that time he had labeled the press "the greatest aggregation of educational foundations," but he had followed up with sharp comments on their power and roles and had demanded better work. Hutchins noted that the numerous and powerful newspaper leaders had paid no attention to the speech, given "before the Depression, before the New Deal, before the Newspaper Guild, before the suburbs, before they charged for newsprint, before the atom, before television." In 1955, he observed, with 800 fewer newspapers in the nation, his voice might not fall on such deaf ears. But, then again, it might, since the reception given to the Commission's 1947 report had been anything but friendly. Editors and publishers showed much antagonism to conclusions such as these:

> If modern society requires great agencies of mass communication, if these considerations become so powerful that they are a threat to democracy, if democracy cannot solve the problem simply by breaking them up—then those agencies must control themselves or be controlled by government. If they are controlled by government, we lose our chief safeguard against totalitarianism—and at the same time take a long step toward it.[4]

Hutchins noted that he had expected to receive the Society's congratulations after the 1947 report was released. Instead, in his opinion, he had been saved from condemnation only by the "unwillingness of your committee to *dignify* me by such action." The president of the society in 1947 accused members of the commission headed by Hutchins, then chancellor of the University of Chicago, of being "left-wing"; and he suggested that, as most of them were professors without experience in the newspaper business, their conclusions ought not to be considered important. Was that gentleman especially angered by the Commission's proposal that the "members of the press engage in vigorous mutual criticism" or by the proposal that an independent agency be created to "appraise and report annually upon the performance of the press"? It could be that he would have ignored or minimized the criticism that minority groups were excluded from "reasonable access to the channels of communication." Perhaps what was disturbing was the Commission's recommendation that lying by the press be investigated, with particular attention to "persistent misrepresentation of the data required for judging public issues." Or perhaps it was the strong statement about the obligations of professionals:

> The socially indispensable functions of criticism and appeal may be as abhorrent to the diffident as they are attractive to the pugnacious, but for neither is the issue one of wish. It is one of obligation—to the community and also to something beyond the community, let us say,

to truth. It is the duty of the scientist to his result and of Socrates to his oracle, but it is equally the duty of every man to his own belief. Because of this duty to what is beyond the state, freedom of speech and press are moral rights which the state must not infringe.[5]

Hutchins was still on the attack in 1955, demanding that the powerful print echelon of the press stop taking what he called the "official line"—that the "international conspiracy" was *the* cause for concern—while freedom was eroding in the nation and the world. He accused the press of being late and timid in attacking Senator Joseph McCarthy, of concentrating public attention on "sinister figures on the Left" while virtually ignoring "fat cats on the Right." He declared the press inept in coverage of how security standards were being generally applied in the federal government and how various agencies (the Justice Department, the Post Office, and so on) were engaged in feverish but questionable activities which mocked or reduced due-process protections. Hutchins ended his address by again calling for a critical press, one that "presents the alternatives the people must know about—telling the people where they are in time and space." [6]

Most astute critics feel that Hutchins has few peers as an irrepressible agitator for press responsibility. The recommendations of his commission of 1947 have weathered well and are obviously pertinent to current affairs; they need amendment from time to time because of what experience shows to be necessary, of course, but the issues themselves remain. In 1967, for example, presumably concerned with press coverage and duties relating to the Vietnam war, and with upheavals changing the civil rights and political party situations, the editors of the *Columbia Journalism Review* recognized the need for a review of the Hutchins report. The thirteen recommendations, as updated by key professional observers, specified that:

. . . constitutional guarantees of the freedom of the press be recognized as including the radio and motion pictures.

. . . government facilitate new ventures in the communications industry. . . . Where concentration is necessary, the government endeavor to see to it that the public gets the benefit.

. . . legislation [be enacted] by which the injured party might obtain

> a retraction or a restatement of the facts by the offender or an opportunity to reply.
> . . . the government, through the media . . . inform the public of the facts with respect to its policies and of the purposes underlying those policies.
> . . . the agencies of mass communication assume the responsibility of financing new, experimental activities in their fields.
> . . . members of the press engage in vigorous mutual criticism.
> . . . nonprofit institutions help supply the variety, quantity and quality of press services required by the American people.
> . . . academic-professional centers of advanced study, research and publication in the field of communications [be created].
> . . . a new and independent agency to appriase and report annually upon the performance of the press [be established].[7]

Perhaps the most interesting aspect of the 1947 effort was that the recommendations came from *outsiders* who tried to change the collective national attitude about the press. The impact of its conclusions has been subtle but steadily growing in importance to many opinion leaders. With each new national crisis, more opinion leaders sense how right Hutchins was to take the lead of a concerned group representing social and philosophical forces. Nevertheless, and despite the Watergate affair, one appreciates why even eminent critics chosen to review the findings in 1967 were still reluctant to give tribute to the Commission. We are just beginning to realize that even the traumas of the early 1970s may not be enough grist for the mill of reform. Incredible as it may seem, Hutchins's warnings of 1930 to the American Society of Newspaper Editors, and all his subsequent demands for reformation of the media, still tend to be ignored by professionals who know how short-lived is the usual public outcry for better press protection against arbitrary and undemocratic government administrators.

There was a widespread sense of euphoria about how well the press had served the nation during the crisis culminating in the resignation of President Nixon. If that euphoria blinds us to the fact that most newspeople did little noteworthy reporting on the national crisis, but merely followed reluctantly on the heels of a few courageous and professionally able leaders, it is sure to blind us to current criticism.

CURRENT CRITICISM OF THE MEDIA

Poor Press Coverage of Stories

The word *crisis* is losing specific meaning because it is attached to tensions ranging from international disturbances to sheerly personal problems. However, it is fair to relate the word to the mass media which, by underplaying or ignoring important stories, may help to create the public misunderstandings so vital to the chemistry of crisis situations. Contributions of the media which enhance democracy must not blind us to extremely valid criticisms which highlight shortcomings.

High on the list of shortcomings criticized is that the *biggest stories* are covered very poorly by the press, and so institutions and processes of government are shortchanged. John Tebbel, going over what the *Columbia Journalism Review* listed as the "least (or worst) covered stories of 1969," tried to moderate the view that the press is *lame,* a word that one dictionary equates with *crippled, weak, infirm, paralyzed,* and *unable to move*. In an appeal for understanding, he points out that "the press is finding it impossible to satisfy competing ideas of what news is and, no matter what it prints, finds at least part of its audience irate because it did not print something else." Regarding the worst-covered stories relating to "Congress, the Defense Department, the police, the courts, state legislatures, local governments, medical care, education, industry, and the media themselves," it is hard to refine the general public's tastes in news, even if the press attempts to go beyond the traditional who-what-where approach. According to Tebbel, presentation of the complexities of a highly debatable subject tends only to confuse or anger the large segments of the public now accustomed to expect simplicities on great and crucial matters.

"Medical care," says Tebbel, "remains one of the great unreported stories in the United States, but again, full reports of this open natural scandal might only frighten and anger readers and add to their disbelief in the media." In the case of stories about the Defense Department, he concludes that "everything that emanates from Defense has a political coloration, so that no matter what the newspaper prints, part of its readership will think it is lying." [8]

One takes no real comfort from Tebbel's cautious and well-founded practical definition of news as "anything that happens." Even though he deplores the fact of government domination over the press, given the fact of overt surrender of public rights to caretakers in government (who may be too unreliable to balance the public interest with their own newly acquired power), he dwells more on the problem than on the solution. With the illustration of former Vice-President Agnew denouncing television as overly powerful, in his speeches of 1969 and subsequently, Tebbel leaves us worried about how the press should handle such a powerful political entrepreneur. We know that the press was obliged to cover Agnew's diatribes against television and the print media because he was a "heartbeat" away from the highest office in the land. We also know that Agnew registered well with the broad segments of the citizenry because his denunciations of the press appeared to be a good verbal hook on which to hang much anger about troubles at home and abroad.

Tebbel offers us insufficient satisfaction, pointing out how much better the press is than it was at the turn of the century regarding treatments of leaders or events. For example, he shows how the reportage of the *Chicago Tribune* on the Spanish-American War would now be described as "blatantly jingoistic, far different from the relatively objective war reporting of today." To be sure, the press has improved, but the demands made upon it have increased out of proportion to that improvement.

In his attack on the press, Agnew was no rather simplistic leader like William Jennings Bryan, who attempted to shape public opinion in his newspaper *The Commoner* or in his occasional forays on radio in its early days. Agnew must be judged in the light of what happened, from the 1930s to the 1960s, in countries controlled by executives who began by intimidating the press and ended up by eliminating the press and all freedom. Tebbel's Spanish-American War based contrast—regardless of the press's proficiencies or deficiencies—refers to a conflict that did not drain resources of the nation and the world. The Vietnam conflict, by contrast, tore at our substance and our souls. The longer it went on, the more the structures of world politics tottered crazily. To imply that we have a better press now is to dismiss the virtually overall acquiescence of the press corps in the years before the Tet offensive of 1968. In judging the quality of coverage of any story, one deals with the total environment. Agnew was seeking something more than fairer treatment of the administration, but the real meaning of his objective is buried in the ruins of the Nixon government. To our regret, we still know much more about the step-by-step military entrapment in the morass known as Vietnam than we know of Agnew's purposes.

Television's Inadequacies

On the whole, the press has thus far protected the Constitution by professional work that enhances its rights and responsibilities. If one looks at some recent tendencies, the generality loses punch. In point, local television stations are frequently observed suffering from advanced news-anemia. It is no secret that many, if not most, local stations broadcasting on VHF frequencies lack sufficient professional staff to cover the news of their own communities. As a result, they cover up by employing personalities who orate the news, in pseudo-Hollywood fashion. The news to them is predominantly what is called "rip and read" materials. UHF television stations, heralded for a while as hopeful instruments of community regeneration in a world becoming more impersonal, are often marginal operations without even the camouflage of personalities to mask news inadequacies.

In our bigger cities, there has been frantic competition between television stations for "anchormen" who have the charm to lure viewers away from the news programs offered by rival stations. With the theatricality thus brought to presentation of the news, the essential talents of reporters as news ferrets tend to be overshadowed, and the glory goes to the announcers rather than to the detectives. There is no necessity in this—just conditioning encouraged by the rating system. It matters little how a station gets viewers, so long as it gets them. The Pied Piper would make a terrific master-of-ceremonies on news programs if he could only be brought to life. The comptrollers of television stations and networks would immediately see his profit-making possibilities. But alas, his drawback: he not only would get the audiences but would disappear with them.

There is some countereffect even in respect to the many stations gone theatrical. Increase of news-related profits has meant *some* new interest in news gathering. In 1974, for example, WFSB in Hartford, Connecticut, not only expanded its evening news program from a half-hour to an hour but also established news bureaus throughout the state. WNBC-TV of New York City, grappling with WABC-TV and WCBS-TV for Nielsen ratings, augmented its news staff by installing a respected longtime newspaperman as executive editor and emphasizing the work of its consumer affairs unit by bringing in Betty Furness as its head.

On the whole, news professionalism is reduced in importance; a new industry has sprung up. News consulting companies offer "show-doctor" advice and "head-hunting" services to aid in the search for anchormen. The cold, hard fact behind this new industry is that "the gain of a single rating point in New York or Los Angeles at 6 P.M. can be worth $500,000 a year in additional revenues." In St. Louis or Baltimore a single rating point increases income by $125,000 a year.

One result of "show-doctor" advice is manifested in the efforts to feature exciting newsfilm and "more sensational police blotter stories." As one organization, Rierson Broadcast Consultants of New York, put it in a report to one of its clients, "It should be the goal of a television news producer to present every story on his program in visual terms." [9]

Coverage of News in New Hampshire

In New Hampshire, where news coverage of subjects of statewide importance is affected by domination of the print media by one newspaper and by the small amount of significant locally produced news available to television viewers, there has been a development which should encourage people in a similar situation elsewhere in the nation.

There is only one statewide newspaper, the *Manchester Union-*

Leader, run by the irascible William Loeb, whom one critic has labeled "a man who is a little to the right of [notorious inquisitor] Torquemada." Two cities, Concord and Keene, have respected local newspapers. Many New Hampshire citizens depend on Boston and New York City newspapers, and television is dominated by Boston stations or network "feeds." On the whole, the state is "a colony of the national media." More than 800,000 inhabitants must turn to Mr. Loeb's newspaper to satisfy much of their requirement for state news. A hopeful sign in an otherwise bleak local news situation is the television program "The State We're In," produced each weeknight by WENH, the public television station at Durham, associated with the University of New Hampshire. Twenty minutes long, each presentation is distinctive in that news and views alternate to the Loeb paper's are offered. The program is a responsible window on the state which reduces the virtual news monopoly. At issue with the Loeb position on a proposed huge oil refinery, WENH helped to cancel out a project it opposed on ecological and other grounds. A side point of considerable import is that this little effort on public television achieved considerable prominence under financial circumstances that discourage individuals and groups who would establish a rival newspaper of statewide significance.[10]

The Haley Study

Sir William Haley, former editor of the *Times* of London, director-general of the British Broadcasting Corporation, and editor-in-chief of *Encyclopaedia Britannica,* was eminently qualified to undertake a sixteen-month-long personal study of television in the United States. At a peak time of war and crisis, as we turned into the 1970s, he found that American television stations and sponsors displayed a singular "lack of enthusiasm" for news documentaries. Mincing no words, he declared an elemental and fundamental fact: "It can have grave consequences for American democracy—unless the news stories are restructured or a corrective is provided in some other way."

What is news to Sir William Haley? First, he says what it is *not:* "It is not a happening." On the contrary, "It is what journalists make of it. It is the sifting, reporting and evaluating of what has happened." [11]

Haley's view is of fundamentally critical importance if we are to reveal politics to the crucial electorate, that taproot of democracy holding fast to our historical heritage against every adversary of democracy. If politics becomes more murky because of half-truths and partial stories based on "happenings," the electorate will lose its hold. For a graphic representation of the collapse of political legitimacy as evidenced in the rise of the Nazis to power, one need only view Alain Resnais's *Night and Fog.* [12] In this modern documentary classic, the French cinematographer leads us down now peaceful weed-covered railroad tracks to architectural monstrosities which began as concentration camps and soon became extermination camps. With his uncanny genius for capturing undeniable truths, Resnais gives evidence of the "grave consequences" resulting from the destruction of the press and all that it stands for—responsible executives, legislators, and courts, and, above all, the principles of constitutional government.

Because of economic pressures which have made intracity competition unattractive or impossible, the majority of American cities have only one newspaper today. Very few important developments are confined to a single locality, be the discussion turned to the price of the family market basket this week or to the social and political costs of racial separation this day or this year. Also, most newspapers copy from the wire services much of what they choose to print about national and international affairs. The consequent inadequate handling of national and international news is becoming a serious problem; and, with its drift toward personalities rather than incisive news coverage, television does little to improve the situation. Indeed, combine a local newspaper's inadequate reports with the superficial news programs of a local television station, and one has a potent brew of unhealthful ingredients. Here are a few of Haley's summary points:

> Visual news values are almost in inverse ratio to real news values. What is most exciting to see is generally the least important to know about.

> Sixteen months of American viewing left me with the conviction that the truth has not yet been realized that even supposedly exciting events by their recurring similarity lose all interest.
>
> Once stories are not tautly edited and lose proportion and significance, the whole idea of news loses significance for the viewer. And the loss of significance is not made up by deep treatment of important and serious news items in documentaries.
>
> TV documentaries have not the skill [of the BBC] at getting at essentials, and the deep probing into them, that British documentaries have. All too often the longer they go on, the more superficial they become.
>
> . . . The opening of the whole world to news, the speed with which it is now communicated to newspapers, radio and TV, the multiplicity of seemingly insoluble national and international dilemmas, have led to a flight from the news by those who, if a nation is to be healthy, should most be following it.
>
> Refuge is sought in generalization. Speculation takes the place of judgment. Today's big news may prove to be tomorrow's trivia.[13]

Spiro Agnew's Complaints

On November 13, 1969, the then Vice President of the United States, Spiro T. Agnew, launched an attack on the three major television networks, utilizing time allotted to him by those same organizations. At the hour in which they usually presented their evening newscasts, he took on the role of Sir Galahad defending ''middle America'' from the nonobjectivity of those he identified as responsible for determining what the news is: ''A small group of men numbering perhaps no more than a dozen anchormen, commentators and executive producers,'' he said, ''settle upon the twenty minutes or so of film and commentary that's to reach the public.'' According to Agnew, these people:

> . . . decide what 40 to 50 million Americans will learn of the day's events in the nation and the world . . . can create national issues overnight . . . can make or break by their coverage and commentary, a moratorium on the war . . . can elevate men from obscurity to national prominence within a week . . . can reward some politicians with national exposure and ignore others. . . .

He went on to characterize the network reporter who "covers a continuing issue—like ABM or civil rights," as an individual who "becomes, in effect, the presiding judge in a national trial by jury."

Expressing especially personal annoyance, Agnew criticized the network commentary that had immediately followed President Nixon's previous week's speech to the nation on the conflict in Vietnam. The reporters, he alleged, had predetermined their views, and the majority were hostile. Referring to Averell Harriman, who was invited by one network to comment directly after Nixon concluded, he called the former ambassador to the Soviet Union, the former governor of New York, the former chief negotiator for the United States at the Paris peace conference with the North Vietnamese, a "failure" who encouraged the country to disregard Nixon's statements. (One cannot help but suspect that Agnew objected to Harriman primarily because he was a lifelong Democrat and longtime leader of his party as well as a spokesman for what the British label the "loyal opposition.") Agnew saw—in the networks, their reporters, and guests such as Harriman—a virtually conspiratorial association, irresponsibly bad-mouthing Nixon.[14]

There has been much well-founded speculation that Agnew was Nixon's chosen spokesman for an administration which considered itself surrounded by enemies. It was not the administration making the mistakes, according to the scripturalists working for Nixon; it was the media—nationally known newspapers and the television networks—which were responsible for the seemingly endless torrent of bad news battering the public.

Agnew found a vast audience eager to hang on to his every word. Despite the tragic conflict in Vietnam, despite Nixon's unfulfilled 1968 campaign claim to have a secret plan for ending the war if he was elected, despite the home-front turmoil spreading from school to school all across the land, despite the rupture of the national economy by reckless spending without consideration of the stripping of the values behind our currency, despite the serious divisions over our foreign policy in general and Vietnam in particular, which went beyond political party loyalties—his speech raised real and false hopes for tens of millions. Perhaps, it was felt, our leadership and policies were not at fault after all! Perhaps it was the messengers who were to blame because they brought bad tidings!

Once launched on his career as grand orator for the administration, Agnew went on and on, elaborating the theme for years, successful in reaching receptive audiences almost to the day he accepted the results of his lawyers' plea bargaining and resigned his high office in disgrace. Like those other great masters of dramatic oratory, William Jennings Bryan and Joseph McCarthy, he became a symbolic spokesman of popular discontent (although, like them, he was ostensibly introspective but inwardly superficial).

Despite the promptings of White House speech writers who had him say what Nixon wanted said, Agnew became important to the general public in his own right. With the first of his speeches on the subject of media monopolies, he managed to tap a deep well of public opinion. Although he was a front man for much arbitrary government, the accusations about press centralization were based on the truth. Again, even if his immediate political goals were self-serving and his tactics generative of much more of the rancor and far-ranging disputation he publicly deplored, there was and is real cause for concern when so much news control rests in the hands of so few.

So Agnew captured favor with "middle America," sometimes called the "silent majority" by Nixon publicists because many people did not know how to deal with the frustrating news that confronted them every time they turned to a network news program. Do not give him too much credit, however, beyond acknowledgment that he shared a concern we all ought to have. Remember that the Agnew who at the time of his first nomination admitted that his name was "not exactly a household word" was elevated from "obscurity to national prominence" in practically no time at all. His ambitions for public acceptance via publicity were well furthered by his anti-press campaign.

Agnew was the beneficiary of organized publicity, as Vice President. For example, shortly after his first inauguration, the U.S. Information Agency rushed to completion an elaborate film for presentation to foreign audiences, to give them a clear account of his personal background and to convey very favorable impressions of his role in American government. It seems that there was at first hardly any useful material available; photographs, some mediocre or poor motion picture film of his inauguration as governor of Maryland, and other limited materials had to be greatly enlarged upon. So using the occasion of a state trip to Asian countries, the USIA sent camera crews to get plenty of good footage. The film opens dramatically as his impressive government airplane is being towed from the hangar at sunrise. By the end of that trip, there was enough film processed to form the basis of USIA imagery of his importance.

One ought not concentrate on Agnew himself to the extent of losing sight of the basic and transcending matter to which he called attention. By denouncing media monopolies he forced public debate on the subject of openness in the press. The subject is above personality, above party, and even above national considerations and, with purely coincidental timing, was central to a letter sent by the dissident Aleksandr Solzhenitsyn to his official persecutors as represented by the Union of Soviet Writers. On the day before Agnew blasted American television, the Nobel Prize-winning author of powerful novels about man's struggle for fundamental dignity lashed out at those who had censured him:

> OPENNESS, honest and complete OPENNESS—that is the first condition of health in all societies. . . . He who does not wish this openness for his fatherland does not wish to purify it of its diseases, but only to drive them inward, there to fester.[15]

LOOKING BEYOND POLITICAL PARTISANSHIP

The Epstein Study

Edward Jay Epstein, in his book *News from Nowhere,* wields no political party's battle-ax in his foray into television news procedures and organization. His painstaking survey of network news programs resulted in some support for the Agnew thesis that something was fundamentally wrong with the news presentations by which most Americans were influenced. In a study begun for a doctoral dissertation at Harvard University in 1968, he sought to answer this question: "To what extent are the directions that large organizations take, whether they are political parties, city governments, business corporations or whatever, determined by pressures to satisfy internal needs rather than by external circumstances or even long-range goals?"

His field study commenced in September 1968 at NBC and continued through 1969, with intensive scrutiny of news operations at the three major networks. The Federal Communications Commission, network-affiliated stations, and peripheral organizations in public relations, advertising, and the law were also found to be fertile sources of information. Epstein interviewed scores of "correspondents, news editors, producers, technicians and network news executives." [16] His conclusions, the products of scholarly research and caution, proved to be just as shocking as Agnew's naturally provocative charges.

He discovered that despite the increasing complexities of news events covered, television journalists with noteworthy formal education are few in number. The road to success is found in climbing the corporate ladder, starting almost at the bottom. In one interesting case cited by Epstein, an ill-trained reporter covering a teachers' strike in New York City reached certain conclusions about the basic reason for the walkout, only to have his interpretation altered by the NBC Evening News producer, who felt that the story should be "played another way." Epstein points out that "the correspondent, since he could not legitimately claim any special expertise in the matter, had to yield to the producer," [17] who had no personal knowledge of the event as it was unfolding, other than provided by the reporter. Should we not entreat network executives to maintain a higher standard of proficiency in and through their hiring practices?

There are two major areas of Epstein's investigation which deserve particular mention here. The first deals with the limitations on the news imposed by the organizational structure and traditions of the networks.

Among the impediments to the goal of the completest story, that dream of all responsible newspeople, is the trend toward *group journalism*. Assignment editors (controlling who goes where, to cover what, and when it will be covered) tend to favor random assignments for each reporter over the alternative of allowing him or her to gain increasing knowledge on individually specialized subjects. Ad hoc assignments, as a routine, send reporters ranging all over the map, both geographically and intellectually. So the general-purpose reporter has emerged as the main strength of network news personnel. There are exceptions, of course. Each network sees to it that key correspondents are assigned for sustained coverage of certain institutions or of evolving major stories. The White House, the Pentagon, the federal courts in Washington, as well as "Watergate," "The Economy," and the peripatetic Henry Kissinger are among the bases for reportage considered so important that individual talents must be kept on to supply the network with all details. On the whole, however, rotation from one story to another is the common experience of network reporters. It has happened that some have had to travel, physically and mentally, and in comparatively short periods of time, from school desegregation, to labor troubles, to train wrecks, to Senate races, to automobiles and pollution, to how much the "market basket" costs in a given place, to whether a certain politician played fast and loose with the funds for a local fire department. When we see correspondents on national television each night, most of us are unaware of what must be a cruel physical strain on them as they are pushed and pulled by assignment editors.

This system of professional news gathering represents a problem with obvious ramifications. It is naïve to believe that the networks are organized well enough or staffed adequately to provide efficient, comprehensive, incisive coverage of the news. One assumes that if they were to revamp the system, their costs would shoot up enormously. But is it not about time that those with the Midas touch—and assuredly television has been an entrepreneurial gold mine—invested more in this vital side of their business? Surely it has been shortchanged long enough, with allocations of personnel and funds spread thin. According to Epstein:

> A six-week analysis of evening news broadcast logs shows that ten correspondents reported 68 percent of the film news stories at NBC, 56 percent at ABC, and 51 percent at CBS (excluding those reported by anchormen).

Another of Epstein's significant findings is that network executives have devised rather set guidelines for the presentation of news, guidelines based upon their own assumptions about viewers' needs and interests. They favor "easily recognized and palpable images." Conflict, they have decided, is more interesting than scenes characterized by placidity; action is more attractive than any other aspect of stories, and *activity* must be stressed. Moreover, according to their guidelines, viewers need a traditional format, and presentations should include a beginning, a middle, and a conclusion. It is not surprising that, with executive axioms such as these forming so much a foundation of national news programming, we as a people are diverted from the import and significance of events to those facets which are photographable and which lead to *punchy* summarizations.

The networks, says Epstein, have gotten into the habit of prestructuring news stories according to categories into which they place them. With viewers conditioned to interpret reality instantly in terms of stereotypes drummed into their heads, the name "California" triggers the image of "The Bizarre Setting." In the period of Epstein's survey:

> Almost all stories about California were depicted as taking place in curious, eccentric and highly unpredictable circumstances. . . . An unpluggable oil leak erupts in someone's patio in Los Angeles; governors on horseback and in cowboy hats ride off into the California sunset at Governor Ronald Reagan's ranch; prisoners commute to outside jobs from San Quentin Prison in a novel experiment, and experience California life styles, while inmates at Folsom Prison furtively build an unflyable helicopter in the machine shops; California adults become heavily involved with war toys. . . .

"Turmoil" is the key to many of the presentations about European events and their background. Stories about mainland China highlight "Pomp and Ceremony"; Vietnam meant "The Mechanical

War.'' Congress is characterized, by virtue of the stories made available, as ''An Investigative Agency.'' ''The Mystique of the President'' and ''The Apocalyptic Battle on the Campuses'' also have been hard-stressed themes.[18]

The Current Situation

On the whole, then, there was substance to Agnew's complaints of 1969, and his use of them for politically partisan purposes is no longer the central matter. Of far greater significance now is the fact that there is still no evidence of progress toward reform of the television news system. No one-dimensional cleverness, no image-theme pseudocreativity by television news managers, has done much to help the citizenry grapple with truly crucial problems. The issue of the general-purpose, Jack-of-all-subjects traveling reporter still underscores the news industry's failure to catch up with the parade of complexities in current history. With the exception of the Watergate story, which grew from 1972 to 1975—and for which a number of journalists with law degrees were made prominent on a day-to-day basis, to clarify highly involved developments—the networks have remained largely aloof to the suggestion that specialized staffs be built.

By election time 1974, the American people, tired as they were by a succession of national political disasters, were nevertheless assured of full coverage of candidates in every state contest. That was a constructive development, in that the networks were fighting the social lethargy and general political apathy of the time, but the populace needed more adequate treatment of other vital stories—the rampaging inflation of the United States economy and its worldwide repercussions; the domestic food-grain shortage and the not unrelated starvation of tens of millions of people in India, Bangladesh, and other places; the worldwide shortage of energy resources, a shortage which evidence indicates will become more pronounced with the geometric population increase in coming years.

Roger Fisher, professor of law at Harvard University, compiled a checklist entitled ''The Public's Right to Know: Functions and Malfunctions of the Media'' for an August 1972 meeting of the Workshop on Government and the Media, of the Aspen Program on

Communications and Society. Fisher listed public affairs functions of the media. Where communications *to* the citizens are concerned, he stressed the functions of watchdog, critic, public forum, and aid to public action. For communications *from* the public, he cited the megaphone role, giving individuals access to a broad public and conveying information and criticism to the government.

Evaluating the performance of newspapers and television, he lamented the concentration on ''hot'' news and the lack of attention to significant subjects which may not have a lead tied to an event today. A measure of Fisher's opinion is his valid assessment of the reportorial role in general. With respect to communications *to* citizens, he noted distortion ''by journalistic traditions overly concerned with political effects; not sufficiently concerned with substance.'' [19]

Douglas Cater, former assistant to President Johnson for health and education, the author of *The Fourth Branch of Government*, and now director of the Aspen Program, offered his colleagues an illustration of how truly circular news management has become. About his boss, President Johnson, he observed:

> Anyone who worked for Lyndon Johnson had a graphic personal case study of a politician whose thought processes were being shaped by the news process. What was on that ''A'' wire, what was being interpreted by Scotty Reston, what was getting out on the CBS news at night—this was the thing he was trying to manage. When President Johnson said he was for a bill passed in Congress, he wasn't just posturing . . . but he was shaped in turn by this whole reporting apparatus. We used to despair when we would be having a serious meeting and he would wander over to the ticker; he would start pulling it and read it all the way back and then he would open the doors and disappear into the bowels of the thing trying to catch the word before it even hit the tape.[20]

RESPONSIBLE OR RECKLESS?

Critics, especially those cautiously framing views about the media and politics, know that there probably is little time left for us to ensure democratic communications for the next generation. One concern is that a meaningful counterattack against the evils of centralization may not take place. In short, critics worry that they are describing a slide rather than a climb. There is no debate on the facts relating to the growth of the media in recent decades. Between 1950 and 1973 newspaper circulation in the United States rose modestly, registering a 17 percent rise to 63,147,000. Magazine circulation for the same period rose significantly, up 64 percent to 241,051,860. Television, in terms of the number of households with sets, has expanded fantastically, to 65,600,000 in 1973. In the period since 1950, television receivers have multiplied by 1582 percent.[21]

A hopeful sign in an otherwise less than sunny picture is that the press corps as a whole is beginning to criticize itself, as if sensing that there is no guarantee of a free tomorrow. Tempered by the political fires of the last decade and more, newspeople are no longer content with slogans such as ''investigative reporting'' and ''advocacy journalism.'' The more profound critics try to see to it that no tribute is given to any reporter who fails to do his or her basic job.[22]

Gone are the days when the news professionals could consider that they somehow stood apart from what they reported about. The height of cowardice is reached whenever a major segment of the press avoids a challenge to any government representative or private pressure group that tries to withhold the truth. Another evasive tactic is evident if the press pretends to be doing its job when, for one reason or another, it allows mostly superfluous information to be disseminated on any given subject. Government officials, often guilty of gross malpractice in public relations, have utilized many disguises to cloak the truth. The news industry not only must strip away those masks from officials but must avoid the same tactics vis-à-vis their own clientele.

In ''The First Amendment on Trial'' (1973), Charles Renbar attributes much evil to what he terms the ''Outlaw Government.'' High officials of the executive branch, he suggests, have been guilty of repressive activity against the press to keep the public in the dark on important subjects. Among his examples is the attempt to withhold the Pentagon Papers from the organs of mass news because of alleged requirements of national security. It turned out that the suppressed material, when made public, was more useful for general education of the American citizenry than it could have been for any enemy; shown clearly were the outlines of many follies, embarrassing to the perpetrators but hardly dangerous in terms of national defense. Renbar is also of the view that the federal charges brought against Dr. Spock were designed to silence dissent during the Vietnam war. Referring to former Vice President Agnew, Renbar deals with his attacks on the press in 1969 and subsequently:

> The police chief threatening the local bookstore is acting no more unlawfully than the Vice-President of the United States threatening whole sections of the press. Each, without warrant in law, attacks rights granted by the First Amendment.[23]

The essential point: if any authoritarian group gains control over the press of the United States, dictating content, the light that nearly died when Hitler conquered most of Europe will finally be snuffed out for all foreseeable time to come. Agnew's fulminations, even in a time of great domestic and international stress, were bearable because they emanated from the mind of an essentially local politician who pretended to be sophisticated and experienced in national matters—local politician makes good and shows off, as it were. As it turned out, the inner circle of the Nixon White House were all at about that level. Nixon permitted the Agnew speeches because he, too, enjoyed playing with fire and seldom rose personally above the slogans of his administration. Suppose for a moment that the attacks on the press had intensified and that the amateurs had been reinforced by shrewd professionals of an inherently anti-democratic stamp. The long and painful, but democratic, process associated with the Watergate case—the Senate investigation, the House of Representatives impeachment hearings, a Presidential resignation, trials in free courtrooms, and so on—might never have happened. There might have been no extrication from arbitrary government.

LEARNING FROM AMERICAN HISTORY

Earlier Days, Same Issues

Learning from the history of struggles to protect freedom of expression in this land, we note that each new generation of Americans has had to renew the political compacts framed in the eighteenth century by theorists, politicians, and common folk. Looking back briefly at early attempts to stifle freedom—and at the protests against that manoeuvering—helps us to put comparatively recent events into perspective. Franklin, Zenger, Livingston, Madison, Kennedy, Nixon, and Agnew are all on the tapestry of history which the next generation will study as it tries to deal with its own perplexing problems. Truth is the basic requirement for responsible republicanism and democracy; and the press constitutes the only efficient instrument able to propagate the ideas, the stories, and the standards relating to governmental respect for and compliance with truth. These generalities lead to an equation of undeniable importance: *Truth* divided by *unconstitutional government* equals *reporting about the destruction of civil liberty* divided by *the collapse of legitimate politics.*

From the beginnings of the United States to the present, there have been waves of assaults upon the free press from those motivated by authoritarian ambitions. The founding fathers of the republic created the First Amendment of the Constitution with more than their ideals in mind: they had the history of such tribulations as were endured by James Franklin, Benjamin's elder brother. He printed in his newspaper, the *New England Courant* (founded in 1721), opinions adverse to those held by important government officials, established church officials, and private personages. In particular, Franklin and his associates published audacious attacks on the theocratic controls exercised in Massachusetts by Increase and Cotton Mather. The Mathers defended themselves in the *Boston Gazette* and, after one especially potent attack, had recourse to the power vested in the elder theocrat to license the press. James Franklin was interviewed by the Council after his publication of charges that the government had not taken proper action against pirates. He was then jailed for a month, and a special committee of the legislature was appointed to deal with him. Finally, the general court decided that the secretary of the province would supervise any and all of his future publications. James avoided the consequences of the order by the subterfuge of making Benjamin, his apprentice, the publisher of the newspaper. Thc authorities saw through it but desisted from renewing the controversy. Increase Mather, obviously displeased, wrote, "I can well remember when the Civil Government would have taken an effectual course to suppress such a cursed libel."[24]

The case of John Peter Zenger, the printer charged with offenses against constituted authority in the province of New York, is well known but bears review. The indictment drawn in 1734 alleged that he "[did] scandalously print and publish . . . a certain false, malicious, seditious, scandalous libel, entitled the *New York Weekly Journal*." The *Journal* replied by attacking the authorities for hiding behind:

> . . . retrenchments made of supposed laws against libelling, [which,] gentlemen, may soon be shown to you and all men to be weak, and to have neither law nor reason for their foundation. [The people of New York City and Province] think, as matters now stand, that their LIBERTIES and PROPERTIES are precarious, and that SLAVERY is like to be intailed on them and their posterity, if some past things are not amended.

Zenger's lawyer, Andrew Hamilton of Philadelphia, one of the leading liberal barristers of the day, did not deny the facts of publication. He turned the case against the government by advocating the duty of freeborn men to fight "corrupt and wicked magistrates" and other officials who misrule the people. Zenger won, resting his case on the truth and the right of the press to publish it.[25]

After winning independence, the nation went through a severe testing of the meaning of the Bill of Rights, when Federalist party leaders took advantage of the difficult international situation and the adverse domestic reactions to French machinations with American envoys during the XYZ affair. They pointed out the fearful dangers of French subversion and military schemes. Republicanism's grip on popular thinking lost considerable ground as an aftermath of the propaganda about real and imagined dreads. While the fire was hot, the Federalist-dominated Congress passed four acts during June and July of 1798, taking aim at the alleged French threat, aliens, anti-Federalist pamphleteers and editors, and criticism of the administration in general. These Alien and Sedition Acts of 1798 bear some resemblance to the acts and attitudes of persons at or close to the White House between 1970 and 1973: there were declarations against conspirators, actual and potential, in both times.

To illustrate, Section 2 of the Sedition Act declared unlawful the writing, printing, uttering, or publishing of any "false, scandalous and malicious writings" against the administration, the President, or the Congress with intent to defame ("bring them into contempt of disrepute; or to excite against them . . . the hatred of the people of the United States"). Libel and sedition were broadly defined to apply to all those who would oppose the administration politically.

The opposition in Congress was spirited, if unsuccessful, at the time. Some thoughts expressed then have lost no validity over time. Edward Livingston of New York warned of the "secret tribunal where jealousy presides—where fear officiates as accuser, and suspicion is the only evidence that is heard." Albert Gallatin of Pennsylvania noted that political writings almost always contain opinions as well as facts.[26]

When the Alien and Sedition Acts were debated in the Virginia legislature, there was much bitterness. Condoning the legislation, General Lee asked, "What honest man would complain of a law which forbids the propagation of malice, slander and falsehood?" Mr. Daniel countered, reasoning that "to combine, conspire, council and advise" were the means that the people used to protect themselves from the "tyranny and oppression of government." [27]

On November 16, 1798, the Kentucky legislature passed a series of resolutions pertaining to the unconstitutionality of the new laws. These resolutions had been drafted by Thomas Jefferson, who labeled them "merely an experiment on the American mind, to see how far it will bear an avowed violation of the Constitution." [28]

Virginia's legislature passed its resolutions, drafted by James

Madison, on December 24, 1798. The tone of the Kentucky-Virginia resolutions reflected the Kentucky view that they were not enforceable or constitutionally binding, because of the inherent violations of freedom of religion, of the press, and of speech. Virginia's legislative assembly decided that "it would mark a reproachful inconsistency and criminal degeneracy, if an indifference were now shown to the palpable violation of the liberties of conscience and of the press." [29]

Although the Federalists managed to win considerable support for their views during a time when emotions ran high in the country (every state north of Maryland rejected the Kentucky-Virginia idea that states could determine constitutionality), they lost power in the election of 1800. With their enforcement of the Alien and Sedition Acts, especially their handling of the Sedition Act, they turned the public against the party. The reaction was based on quick perception of the issue and of the party's undemocratic motives, and not necessarily on the actual cases brought by the government. Under the terms of the Sedition Act, about twenty-five individuals were accused. About fifteen were indicted, and some ten cases came to trial.[30]

Kennedy and the Press

It all seemed so refreshing. Telecasting the inauguration of John Fitzgerald Kennedy in 1961, the national networks caught all the pageantry connected with the handing of authority to the first President born in this century. A hard campaign, in which a noxious religious issue had been exorcised, ended in a clarion call for new attention to the plights of the deprived. Soon afterward, we warmed to the notion that it was a national obligation to land men on the moon by the end of the decade.

The change in leadership appeared to represent more than a mere exchange of the old hero, Eisenhower, for the younger Kennedy. There was an aura of regeneration of national spirits. The press corps was captivated, and all media featured the doings and sayings of the new leadership. Camelot, that mythical center of gallantry and courage populated by King Arthur and his court, had come alive on Broadway with an immensely popular show. The new First Family were made out to be the leaders of a new Camelot, by a large segment of the press turned worshipful or cannily capable of responding to what they considered public fascination.

On Inauguration Day there was great national satisfaction with the democratic utility of the media, especially television. Immense television audiences had already watched four debates between Nixon and Kennedy. It is estimated that 75 million people watched the first debate on September 26, 1960.[31]

The eloquence of John Kennedy's inaugural address still rings, even though the messages sound overly confident and somewhat simplistic after all that has happened. Whatever declaration we use to tingle the mind and recall those days, the rousing nature comes straight through time and space: "Let every nation know, whether it wishes us well or ill, that we shall pay any price, bear any burden, meet any hardship, support any friend, oppose any foe, to assure the survival and the success of liberty." [32]

Press responses were in keeping with the new electricity in politics. Kennedy revitalized the press conferences, appearing regularly on *live* broadcasts and telecasts; about 65 million people saw the first conference held by the leader of the "New Frontier" administration. By dint of personality and intelligence, the President adroitly used television to add luster to his political plans. Through televised press conferences, televised chats with key figures of the Washington press corps, and other means, he became the embodiment of what the nation longed to see in its national leaders. Ben H. Bagdikian, reviewing the situation, credits Kennedy with insight into the ways that television and the print media interrelate. He says,

> Television gave the President direct access to the public. For any Democratic President who knows that at election time or in a national fight with the business community the majority of newspapers will be against him, this was of enormous importance and John Kennedy knew it.[33]

Kennedy tactfully manipulated press interest, naturally taking advantage of the admiration of reporters—and the majority did admire him. Administratively, his short time in office was critically short of substantive achievements. Politically, it was the image of the creative, visionary, and courageous government to which Americans are drawn. Its failures vis-à-vis the mass media were modest when compared with what came after, but they were failures nevertheless.

One shortcoming was in Kennedy's initial tendency to overplay his oratorical talent on occasions when actions spoke louder. What worked with the U.S. Steel Corporation to "jawbone" prices down did not work when deep social issues were involved. One illustration is taken from the events of the autumn of 1962, at the University of Mississippi's Oxford campus. James Meredith, sustained by a federal court order, wanted to enroll, to become the first black student. The twenty-nine-year-old Air Force veteran became *the* issue goading racists, and a stridently segregationist governor added fuel to the fire. Even the 320 United States marshals on the scene failed to cool the tempers of the mobs yelling "Keep Mississippi white" and "Give us the nigger."

The "Battle of Oxford" raged one long night. The groups opposed to desegregation found some professional leadership, insofar as fighting tactics were concerned, in the person of former Major General Edwin A. Walker. Ironically, he had been in charge of federal troops at Little Rock, Arkansas, in the school desegregation confrontations five years earlier.

It was a close thing, with the marshals holding out by the skin of their teeth. A reporter for Agence France Presse was killed. In the morning his body was found. He had been shot by "person or persons unknown."

While the fearful struggle took place, Kennedy tried reason, via a nationally televised speech. Here is a not unrepresentative passage:

> This Nation is proud of the many instances in which governors, educators and everyday citizens from the South have shown to the world the gains that can be made by persuasion and goodwill in a society ruled by law.

In May 1963, the President, discussing a bad situation in Birmingham, Alabama, remarked that it was no time for a "fireside speech" on civil rights. The time for moral testaments, he observed, is "before the disasters come and not afterwards."

Sometimes the media served national security goals as the Kennedy administration needed those goals served. A prime example is his speech to the nation on October 22, 1962, shortly after the Oxford incident. In that speech Kennedy roused the public to a crisis which had been some days at the crucial level but kept secret. The third sentence of that talk about the "Soviet military build-up" in Cuba was, "Within the past week, unmistakable evidence has established the fact that a series of offensive missile sites is now in preparation on that imprisoned island." After telling the people that a quarantine had been imposed on Cuba to keep out all offensive military equipment and the like, he said, "Any nuclear missile launched from Cuba against any nation in the Western Hemisphere [would be considered] an attack by the Soviet Union on the United States." [34]

There is no doubt that Kennedy was politically effective on the Cuba matter, to the extreme. Most critics hold that it was a marriage of media and national security when anything less would not do. Others, more sober, recognize that, to take a phrase from General Hugh Johnson's blast at a Franklin Roosevelt action many years earlier, the administration was "shooting craps with destiny."

The issue behind the event was news management. Arthur Sylvester, Assistant Secretary of Defense for Public Affairs, in a 1962 speech to the Air Force Association, was quoted thusly by a reporter:

> Today in the cold war, the whole problem of information, how it is used and when it is used, when it is released, becomes a very vital weapon. . . . Determination of releasing, withholding in this sphere must lie not down the line, or with me, but has to be geared into what the man who goes before the people every four years to submit his stewardship, no matter who he may be, must be in line with what he and his top advisors are doing.[35]

The Pernicious Element: Nixon's Debacle

John Kennedy's associates advocated skills in public relations and were interested in the idea of news management. By and large, their political approach was to favor certain newspeople or organs of the press when opportunities arose for influencing public opinion. Kennedy had great difficulty in translating his hopes into administrative realities but, when it came to the press and press freedom, his heart was in the right place. With notable exceptions, the press corps thought of him as an ally. Credibility gaps existed but were not a reportorial preoccupation.

Somewhat more than a decade later, we were all stunned by the collapse of the Nixon government. *Watergate* is accepted in the language as a reference to corruption. Before we examine the significant shifts in our history which may have produced the setting for Watergate, a brief review is in order.

The evils of the Watergate scandals had an intellectual base curiously similar to the foundation of excuses which supported the Alien and Sedition Acts. It seems that the inner core of President Nixon's advisers in the White House considered themselves surrounded by *enemies*. Their emotional fixation made them determined to neutralize those enemies by the expedient of managing the news media. Such management, utilizing various forms of pressure, would stop the endless stream of distortions they saw in the press reports. It is not clear whether his close advisers duped the President or whether they were merely clever at meeting what they perceived to be his ego requirements. What matters most is that Nixon's immediate staff (tagged the Palace Guard by the press) was paranoid on questions relating to public opinion and blamed the press and not themselves when their policies went wrong. Richard Nixon is revealed by the historical record to have had a classic persecution complex. He and his associates attempted the manipulation of the federal bureaucracy, especially the agencies for law enforcement and state security, in order to get them behind the drive against the hobgoblins they feared.

Taking office in wartime—though the administration kept the

people as insulated from the conflict as possible by disregarding suggestions for appropriate taxes, controls on business and labor, and so on—Nixon adopted a policy of forcing the North Vietnamese to yield, no matter how long it took. Meanwhile, the press presented ever more complete coverage of the war. After the Tet offensive of 1968, even the most dogged supporters of the government line were disillusioned and showed it by stepping up sharp criticism of President Johnson.

It is interesting that the fall of Richard Nixon and his cronies was triggered by support of criminal activities. After the breaking and entering of the Democratic National Headquarters, they tried to shelter the actual perpetrators under the cover of national security needs of the nation. So much power was in their hands that it is to be doubted that enough evidence would have been gathered to cleanse the White House if they had not been pernicious and stupid.

Despite every blocking tactic used by President Nixon and his immediate associates, *the* tape recordings finally fell into the control of the federal judiciary. And there the true nature of all conversations relative to the crimes became clear. Most of those visitors who talked with Nixon had been unaware that their remarks were being recorded. At one point, before the physical delivery of the famous tapes was forced, Nixon had a version of the bulk of the talks prepared as a transcript for the public. He took to the national radio and television airwaves to explain that all pertinent material was included. Later it was discovered that the transcript he provided did not prove to be identical with the tapes held by the courts and by Congressional committees, especially when absolutely crucial sections of conversation could have documented intent and instructions precisely.

On television and radio, Nixon (April 29, 1974) said, "To anyone who reads his way through this mass of materials I have provided, it will be totally, abundantly clear that as far as the President's role with regard to Watergate is concerned, the entire story is here." [36] It was not! The next day, the President's lawyer, Mr. James D. St. Clair (himself then not privy to all that was on the original tapes), provided the House of Representatives Judiciary

Committee with a 50-page legal argument bearing on the 1308 pages of edited transcripts:

> In all of the thousands of words spoken, even though they are unclear and ambiguous, not once does it appear that the President of the United States was engaged in a criminal plot to obstruct justice.[37]

The United States Supreme Court, on July 24, 1974, ordered the President to surrender the original tapes containing potential evidence for use in the trials, impending and under way, of his former subordinates.

Between July 27 and July 30, thirty-eight members of the House Judiciary Committee voted, by substantial majorities, approval of three articles of impeachment. Nixon was charged with abusing his constitutional authority and violating his oath of office to uphold and defend the laws of the land by (1) following a course of conduct designed to obstruct justice, (2) engaging in conduct violating the constitutional rights of citizens, and (3) attempting to impede the impeachment process by defying legal subpoenas for evidence.[38]

On August 9, 1974, President Nixon resigned in disgrace. Less than two years after the electoral landslide of 1972, which appeared to signify his stature with the people, he was forced out by fear of impeachment articles which would probably have led to his trial and condemnation through the ultimate judicial process applicable to Presidents of the United States. Only ten months earlier, Vice President Agnew had resigned after being implicated in illegal dealings with individuals more interested in payoffs for business favors than in matters of government.

Newspapers, magazines, television networks and stations, and radio networks and stations played as much of a role in Nixon's downfall as did the long investigation of Watergate by the Senate committee chaired by Senator Sam Ervin (spring and summer of 1973) or the House impeachment committee that reviewed the evidence under the chairmanship of Congressman Peter Rodino. There was a never-ending cascade of coverage from a free press alerted by fire bells rung because of foul odors and smoke from the White

House. The great "coverup" tantalized the press as much as what was finally opened to public scrutiny. Rising to the occasion as never before, the media were eager to carry out their most serious political duties of protecting the democratic spirit and constitutional procedures of the republic.

It is fashionable now to certify individual newspapers for commendation in this process of rectification. The *Washington Post* and the *New York Times* certainly qualify. So do the national newsmagazines *Time* and *Newsweek*. And the television and radio networks, NBC, ABC, and CBS, also deserve accolades for bringing the daily ramifications of Watergate so fully to their audiences. However, it is not essentially good to restrict praise to certain leaders. The American press as a whole did its job. Even in diehard conservative areas of the nation, identified as strongholds of the Republican party, local papers carried more and more of the facts. Some wanted to, and eventually all were obliged to, because in a time of electronic mass media and national press organizations there is no way to keep intelligent and free people in the dark.[39]

Even toward the end of his Presidential trail, Richard Nixon failed to perceive the nature of the controversy. He persisted in the false hope that the media could be managed, could be turned from the Watergate catastrophe to coverage of the image he preferred to see of himself—the chief executive of a great power, ready and eager to be revered for his work for peace.

At the end, the media dutifully covered all aspects of what was called a "triumphant" Middle Eastern trip. In the course of seven days in mid-June of 1974, only weeks from his last stand on Pennsylvania Avenue, Nixon visited four Arab nations and Israel. His tour of Egypt was particularly interesting because of the adulation drummed up in that country. President Sadat understood American power more than American political realities. So, it seems, did President Nixon.

By F. D. Fitzgerald. © 1973 by The New York Times Company. Reprinted by permission.

POPULAR PARTICIPATION AND MEDIA

4

ACCESS

Access to information and participation in mass media enterprises which shape public opinion and direct much private thought must be democratically available to individuals and to groups.

Access has special meanings for radio and television license holders. For example, under the terms of the Federal Campaign Act of 1971, "reasonable amounts of time for the use of a broadcasting station by a legally qualified candidate for Federal elective office on behalf of his candidacy" must be provided. Licensees must also take care to respect the Fairness Doctrine of the Communications Act of 1934 (as amended). As Section 315 of that legislation stipulates,

> If any licensee shall permit any person who is a legally qualified candidate for any public office to use a broadcasting station, he shall afford equal opportunities to all other such candidates for that office in the use of such broadcasting station.

Licensees are not permitted to censor materials broadcast under authority of Section 315. Specifically excluded from the mandate are appearances by such office seekers on newscasts, news interviews, news documentaries, and on-the-spot productions covering "bona fide news events."

From 1962 until late 1975, the interpretation of what constituted bona fide news events was based on two cases in which the Federal Communications Commission had ruled that important political debates were not to be so identified. Affected were debates between the two major candidates for governor of Michigan (broadcast by WJR, Detroit, The Goodwill Station, Inc.) and between Governor Pat Brown of California and his challenger, Richard M. Nixon, before the annual convention of United Press International (National Broadcasting Company). On September 25, 1975, the FCC overruled itself on the 1962 cases and on a 1964 decision against the CBS network. In 1964, the Commission ruled that Presidential press conferences and those of any nonincumbent presidential candidate would not qualify for exemption. Under the 1975 interpretation, press conferences of national, state, and local officeholders and political debates may be exempted from Section 315 if they are newsworthy and given on-the-spot coverage. Implied is much greater FCC acceptance of news judgments by broadcasting industry professionals.

In regard to demands from private persons that they be given the right to reply to personal or political attacks delivered against them on the broadcasting airwaves, the U.S. Supreme Court has broadly extended the coverage of the First Amendment of the Constitution. Freedom of the press, since the Red Lion case of 1969, has involved new legal duties on the part of broadcasters. The decision includes the important admonition, "It is the right of the public to receive suitable access to social, political, esthetic, moral and other ideas and experiences which is critical here." The Court clearly indicated that freedom of petition and of speech must be enhanced by the media. Not insignificant was the finding that reply time could be obtained free of charge from the broadcast licensees.[1]

Access relates also to the many forums for community opinions provided by radio and television stations. To illustrate, in the greater Boston area, the UHF facility of educational station

WGBH-TV offers "Catch 44." Week by week, representatives of idea and interest groups present their opinions and talk about activities which they feel to be of broad interest. The station publicizes its desire to make program time available to responsible community organizations. Using another format, commercial television station WNAC sponsors "Mass Reaction." Each week, staff personalities, primarily from the news department, play host to a studio audience and attempt to stimulate presentation of views on a controversial topic. Each radio and television station in the Boston area has its own approach to the access requirements.

These novel programs are interesting but not startlingly impressive, insofar as many critics of the electronic mass media are concerned. They feel that public participation in all creative and administrative processes is absolutely necessary if the mass media are to be truly constructive opinion-shaping instruments.

GOVERNMENT AND OPINION MANIPULATION

Avery Leiserson, a noted political scientist, penetrated the media mysteries of the near future in his 1953 summation on the theories of how political opinions are formed. After recognizing the emphasis on constitutional lawyers and political analysts (who focus on judicial sanctions on the media, administrative rule setting, and questions bearing upon public standards of security and morals), he shifted attention to equally grave problems. Only too well do we appreciate how Leiserson foresaw the battle lines that emerged in the next two decades when he posed these questions:

> How shall the media permit themselves to be utilized by the President, Congress, the military agencies, the State Department, or the information services of the great civilian departments, in return for improved access to news and a recognition of governmental responsibility for advising and shaping, as well as informing public opinion? How should the representatives of the mass media secure access to authoritative sources of program formulation without losing vital elements of independence necessary for public perspective and understanding? [2]

BLACK FRUSTRATION AND THE ALABAMA EDUCATIONAL TELEVISION COMMISSION

Leiserson's warnings about federal government-inspired damage to the media was coincidental with the beginnings of a bizarre case in Alabama which shows that the threats to a free press are not restricted by the administrative level of official work.

The facts are plain. Black citizens of Alabama have endured a very long period of frustration because the state-run educational television stations were vehicles of discrimination in terms of programming and administration. The question posed by circumstances unfolding since 1953 is: If government creates and then manipulates media undemocratically, is there swift and adequate remedy against official mischief? Looking back from a decision of the Federal Communications Commission of December 17, 1974, involving the Alabama Educational Television Commission, one comes across fundamental weakness in the administrative and adjudicative processes.

The AETC was set up by the Alabama state legislature in 1953 and, at this writing, is the licensee for all state-owned and -operated noncommercial television stations. It has ultimate control over all program operations of eight educational television stations and has applied for a license for a ninth. The AETC monopolizes educational television in the state at this time, with outlets in Montgomery, Birmingham, Mt. Chesha State Park, Dozier, Mobile, Florence, Louisville, and Huntsville (a construction permit was applied for to build in Demopolis).

On the face of things, a heavy responsibility devolves on the state, for programming in the best interests of all its people.

Informal complaints were received by the FCC about racial discrimination in programming and employment practices of the AETC during the license term of 1967–1970, from the Reverend Eugene Farrell, S.S.J., Linda Edwards, and Steven Suitts. In 1970, the FCC responded to the complaints by concluding that "there is no substantial problem," and it renewed the AETC licenses. The original petitioners, however, joined by Anthony Brown and William Wright, filed new petitions, and the FCC agreed to review the situation. As late as August 22, 1973, in a proceeding before an administrative law judge, an "initial decision" recommended renewal of the licenses. Curiously, the FCC, in its December 1974 decision, hedged on the 1973 recommendation. It affirmed the opinion but effectively nullified it because of new evidence.

Keep in mind the fact that approximately 30 percent of the citizens of Alabama are black. Also concentrate on the charges against the AETC. Complainants alleged the following: racial discrimination in overall programming and in employment policies and practices; and failure to develop programming to meet the needs of the state's people.

The administrative law judge concluded that, despite the substantial amount of black-oriented programming available from National Educational Television, for example, "AETC elected to broadcast virtually none of these programs." Evidence before the FCC traced a dismal record in the area of "integrated programming" (requiring the appearance of at least one black person per program). The study of sample weeks between October 1, 1967, and January 10, 1970, disclosed the following proportions of integration, in percentages:

October 1–7, 1967	0.7
June 30 to July 6, 1968	1.5
November 10–16, 1968	0.7
December 14–20, 1969	8.2
January 4–10, 1970	12.7

In the FCC text on this case we find that despite the rise in the level of integrated programming toward the end of the 1967–1970 license period, "this additional integrated programming, which AETC acknowledges includes all the further integrated programs revealed in its records—amounts to only 46 hours and 45 minutes of programming, or some 0.5% of the over 10,000 hours of programming broadcast during the license term." With regard to black employment:

> There were no black AETC Commissioners, no black AETC professional staff, and no blacks on the Program Board during the license

> period. The production centers were all located at predominantly white institutions and the record establishes that there was no significant black involvement in the preparation of programming at those production centers. . . . There were no blacks on the Curriculum Committee, an organization created by SDE for the purpose of planning and coordinating instructional programming. . . . No integrated in-school programming was produced locally.

After reviewing the "history of disservice," the majority of FCC commissioners found that renewal was not in the public interest, and the AETC's applications were denied.

In another twist of this drawn-out case, however, the FCC decided that the AETC could continue to operate its stations until new licensees emerged from a process which would permit potential licensees to file until April 1, 1975. Moreover, the AETC, provided it takes positive steps to eliminate its previous discriminatory practices and policies, thereby demonstrating that "it has the capacity to change its ways" and "possesses the requisite character qualifications to be a Commission licensee," is eligible to file applications anew on an equal footing with any other applicants.[3]

So service is continued on Alabama's educational network, and steps have finally been taken to overcome an outmoded and outrageous pattern of discrimination, so flagrantly maintained by the AETC for years. About a decade after the initiation of complaints, true reform is in the air. The administrative-judicial process must be based on prudence and caution. One wonders if it is not sometimes so bureaucratic as to constitute an effective barrier to constructive social change!

If access to the media is to be democratic, the procedure of regulation and adjudication must be made to meet the needs of the times. The majority and the many minorities on any given issue, at any given time, must feel that an appeal to government can bring reasonable and constitutional results before another generation of Americans is harmed by media manipulations counter to the public interest. In the case at hand, the government itself was implicated in discrimination. All citizens were denied proper access to quality television programming. The black citizens suffered only the most obvious burdens. Who knows the social costs paid by the majority? Who knows how much harm was worked by an agency of Alabama upon the psyche of the community? Who knows how much the approaches taken by the administrative law judge and by the FCC have contributed to the constitutional confusion?

HOW MUCH PARTICIPATION?

Jerome A. Barron, in his provocative book *Freedom of the Press for Whom?* (1973), notes that private corporate censorship by broadcasting executives has been defended as a barrier against obscenity. He argues that "the case for access must not be lost because a right of access could be abused."

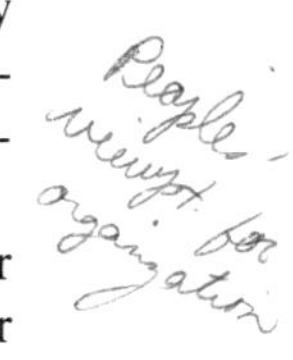

He also criticizes the FCC for failing to make regulations clearer to the industry and for not delineating restrictions in such a manner as to allow both bland and avant-garde fare, according to the needs and maturity of divergent audiences.

Barron is particularly interested in the areas where access and obscenity issues "can sometimes coalesce." Obscenity is one problem for which, he says, "no real effort has been made [by the FCC] to enunciate clear standards."

Barron criticizes the FCC for imposing a $100 fine on WUHY-FM in Philadelphia because the January 4, 1970, presentation of its weekly program "Cycle II" included an interview replete with "the two most famous four-letter Anglo-Saxon profanities." The objection of FCC Commissioners Bartley, Lee, and Wells to the "indecent" programming, observes Barron, indicates that the three probably do not attend the contemporary theater frequently.

The FCC's condemnation of indecency on the airwaves did not extend to a poem read on the "Julius Lester Show" on radio station WBAI in New York, December 26, 1968. A black teacher of history at a junior high school in New York delivered the sentiments composed by a young black poet. In the context of educational confrontations then taking place in the Ocean Hill–Brownsville area of Brooklyn, between some leaders of the local black community and the leaders of the United Federation of Teachers, the poem was more than a flight of artistic license. The UFT leader, Albert Shanker, is of the Jewish faith, as were many members of the organization. The poem, entitled "Anti-Semitism," contains the lines "You pale faced Jew boy—I wish you were dead" and "I got a scoop on you—yea, you gonna die."

The UFT complaint to the FCC asked that that program and one on January 23, 1969, be investigated. On the second program the issue of anti-Semitism was talked about. A guest on the program said, among other remarks, "Hitler didn't make enough lampshades out of them." The host replied that it would be a "dead end street if we get too involved in that hate thing."

Responding to the UFT complaint, the FCC "refused to make any investigation or to take any action in the case." Freedom of speech has its price. When is the cost too high? Given the situation, the refusal of the FCC to investigate is puzzling.

This very touchy area of media access involving personal or group hatred is even less clearly dealt with officially than obscenity, despite the U.S. Supreme Court's decision in *Beauharnais v. Illinois* (1952) upholding a 1917 Illinois statute against group defamation. The state law provided for criminal penalties for publications which assigned "depravity, criminality or lack of virtue" to any class. Prohibited, in addition, were exposure of persons "of any race, color, creed, or religion to contempt, derision or obloquy," and publications "productive of breach of peace or riots." [4]

That decision and many other rulings of state and federal courts are clearly not definitive, because generalized verdicts are hard to apply when the differences between cases are so often more significant than the similarities. Moreover, it is in the American constitutional tradition to shy away from declarations about the Bill of Rights which gloss over basic problems. Case by case, the judiciary tries to be both fair *and* wise. The administrators empowered to deal with electronic mass media have to strike an even more delicate balance: they are obliged to be fair and *practical,* in accord with the law.

How much media-sponsored pluralism can we stand and still retain the spirit of free expression in the land? How can we restrain unbridled hatred? How much defamation is tolerable, and indeed desirable, if repression of personal and group bitterness is not to be enforced by government? How far may the print press go? How far should the presently *conservative* electronic mass media go in order to represent major and minor ideas and ideologies, expressed in personal speech or in the popular dramatic formats? Should the *conservative* media, which dispense violence as routine fare but are timid about offering political radicalism or sexual adventurism or plain nudity, become wholesale purveyors of confrontation?

Can we survive the onrush of groups given full access to the media, with a government sensitive to majority norms and minority needs? Participatory access, for media managers, raises more questions than there are answers for. All responsible citizens who appreciate history know the dangers and, more importantly, the stupidity of repression. Such citizens also fathom the powers of purveyors of fear and suspicion.

Television has reached the stage in the United States where the Archie Bunker character of the very successful "All in the Family" series has been more imitated than complained about. Bunker is portrayed as a lovable white bigot; a racial and religious neanderthal who is also a good family man, surrounded by his forgiving wife and his liberal daughter and son-in-law. As a result of the success of that show, a plethora of similar presentations have recently been molded against the original theme. Black actors have found new territory to stake out as equally lovable bigots denouncing "whitey."

Behind the simplistic, relatively harmless Bunker character is the potential for more dangerous stuff. Suppose television fell into the hands of authorities who liked the basic premise of bigotry, as a forerunner of racial and religious war. They could remove the good-humored covering and unveil the real beast.

At this time, the television medium is so regulated by the industry's censors that we do not know how much media-disseminated personal privilege we could tolerate while claiming to be a constantly maturing people. Day by day we must endeavor to be more democratic and adventurous on this matter of access. Formula radio and television programming constitutes an effective clamp on many expressions of popular will. However, it is also a practical barrier to social misadventures perpetrated through the mass media.

There is much social and political uncertainty, even in the minds of reasonable people who desire that as many voices as possible be heard. Too many good suggestions and reasonable protests have been drowned out already. Still, haunting us all, as we contemplate alternative evolutionary advances for the mass media, are the specters called suspicion, fear, and hatred, buried not too deeply in the human psyche.

For that reason, access for people espousing certain noxious ideas is not necessarily in the interests of most citizens. Access is always conditional, and some conditions are established to enhance the safety of the majority. Nevertheless, a radical minority on one person may be right on an important issue, even if opposed by all the rest of society. For all that, full access may be a genie which democratic people must chase but would fear to catch.

Access to the media is demanded by groups who are emotionally driven to present their political and social views. Frequently, such groups are not satisfied with the regulated forums made available to them. More frequently, their leaders denounce news reports which they contend are unfair or distorted.

The mass media, print and electronic, have had a hard time handling real-life situations in which bigotry and frustration are crucial factors. The national school busing controversy is an issue in point, and Boston is a place in point. Many on both sides of the issue have been angered by newspaper and television reports about the seesaw confrontations between ethnic communities favoring or opposing busing as a means of desegregating public schools. Television has been accused of accentuating certain aspects of the unfolding story, and the local newspapers have been accused of outright bias.

In 1974, the media attempted to take steps to avoid reportage that in any way might exacerbate an already excitable situation, especially in white-dominated South Boston and in the black community of Roxbury. Prior to the opening of public schools in September 1974, a Boston community Media Council was formed. Meeting first on July 10, twenty-six media representatives and twenty-six "minority community"' (mostly black) representatives planned a media and community information program to defuse the tension. Subsequently, the television crews, network and local, tried to comply with Police Commissioner DiGrazia's request to "maintain as low a profile as possible."

Thirty-second public service spot announcements were prepared for the local television and radio stations. Featured were sports heroes, politicians, and community leaders, all of whom spoke for compliance with the law and made it clear that they had confidence in the ability of the good people of Boston to work out their problems in a democratic way. Talent and services donated to the campaign were estimated in excess of a half million dollars.

Despite this unprecedented attempt to reduce emotion, however, the opening of Boston's schools marked the beginning of the most bitterly troubled period in the social history of the two communities chiefly involved with each other on the ramifications of desegregation. The organized media attempt to take no overt position, and thus to be fair, actually tended to sugarcoat all appropriate media coverage with advertisements for goodwill. By backing away from the deeper issues, the media actually might have created untruths about the gravity of the situation.

Legions of reporters from national and local media were there on September 12, 1974, to see how the public school system of Boston would react to the new desegregation busing ordered by federal courts. For months afterward, indeed for the whole of the school year of 1974–1975, reporters covered the unfolding story of strife and bitterness, which made the front pages of the world's newspapers.[5]

CENSORSHIP: THE CONSTANT FACTOR

Government censorship is tolerable in the worst of times, but even then its existence is a threat to the democracy supposedly protected.

Worse than imposed censorship which vests press rights in government is that self-censorship of the press which is a by-product of government intimidation. All censorship which curtails responsible freedom of the press must be fought with that mighty weapon, common sense, in the courts, in the legislatures, and in executive offices.

As has been pointed out, the fight against the censors in this land is older than our nation, and the cause of freedom has rallied the intellectual leaders of each successive generation of Americans. It is the democrat's belief that the wisdom of the national constituency is far more consequential than the conclusions of any one man or clique. With that in mind, democrats have been constantly opposed to capricious censorship.

Even when censorship was necessary during World Wars I and II, it was limited to matters of national security, so far as was possible in times of international chaos. Only twelve days after President Wilson asked the Congress for a declaration of war against Germany, he created a Committee on Public Information, with the newspaperman George Creel as director. Its task was to inform the people, while consistently maintaining national security. The censorship teeth of the government were sharpened by honing from the Censorship Board; the Executive Order of April 28, 1917, on international cable and telegraph lines; the Espionage Act of June 15, 1917; the Trading-with-the-Enemy Act of October 6, 1917; and the Sedition Act of May 16, 1918.

The mood of the Committee, as set by Creel, reflected not so much the idea of restriction as the desire to publicize war aims and war news. As he said, "The work of the Committee was so distinctly in the nature of an advertising campaign, though shot through and through with an evangelical quality, that we turned almost instinctively to the advertising profession for advice and assistance." [6] Under his guidance, the CPI quickly acquired 250 paid employees, 5000 volunteer writers and artists, and 20,000 public speakers. Its news and general publicity campaigns, utilizing the print press, moving pictures, public speakers, advertising, and the like, constituted the most notable governmental effort of information dissemination to that date.[7] Despite the emphasis on propaganda useful to the government, there was a determined effort to maintain cooperation between the CPI and the private-sector editors and publishers. For example, newsmen were asked to look at stories and use their good judgment to delete "matters which obviously must not be mentioned in print" and "matters of a doubtful nature which should not be given publicity until submitted to and passed by the Committee." [8]

Elmer Davis, the eminent newsman who was named director of the Office of War Information in 1942, was careful to respect the intent of President Franklin D. Roosevelt's Executive Order 9182, which recognized "the right of the American people and of all other peoples opposing the Axis Aggressors *to be truthfully informed*." [9]

The CPI and the OWI were adjuncts of the military effort and still adhered to the canons of press freedom whenever possible. On the other hand, there has been a countervailing force in American life in times of both war and peace. Behind it are those addicted to a mental narcotic which makes the idea of a tightly fettered press an attractive illusion. In addition, times of national stress create opportunity for those who hate. After World War I, there was much grist for that hate threshed out of the paranoia about *sedition* and *subversion* allegedly eating away the timbers of the ship of state.

That paranoia gripped U.S. Attorney General Alexander Palmer, who conducted raids on aliens in 1920. It was evident in the expulsion of a Socialist from the House of Representatives in 1917, and in his later trial and conviction under the Espionage Act (although the conviction was reversed in 1921). It was vital in the environment which stimulated the ousting of five members of the Socialist party from the New York State Assembly in 1920; and in the investigation into the *loyalty* of government employees, down to schoolteachers and the like, by the Lusk Committee of New York State's legislature in 1919 and 1920.[10]

That wise man Alfred E. Smith was governor of New York when the Lusk Committee prepared legislation to "detect revolutionary conspirators, to test the loyalty of teachers and regulate the registra-

Extraordinary voluntary censorship during WW II.

tion of schools and school courses with the object of preventing the young from being corrupted by reds and radicals.'' He vetoed the bills one at a time. His opinion is summed up in a portion of his veto of the act aimed at licensing and supervising schools and school courses. We should keep his pungent conclusions in mind as we study the attempts to restrict access to ideas in our day.

> Its avowed purpose is to safeguard the institutions and traditions of this country. In effect, it strikes at the very foundation of one of the most cardinal institutions of our nation—the fundamental right of the people to enjoy full liberty in the domain of idea and speech. To this fundamental right there is and can be under our system of government but one limitation, namely, that the law of the land shall not be transgressed, and there is abundant statute law prohibiting the abuse of free speech. It is unthinkable that in a representative democracy there should be delegated to any body of men the absolute power to prohibit the teaching of any subject of which it may disapprove.[11]

Something of that same paranoia was inherent in the restrictions placed upon the right of American citizens, including the members of the press corps, to travel to mainland China and Cuba in the 1950s and 1960s, particularly during the period of John Foster Dulles's stewardship over the Department of State. Very little came from the federal courts to counter administrative restrictions. In *Kent v. Dulles,* decided by the Supreme Court on June 16, 1958, the majority (5 to 4) found for the deprived citizens. The majority quoted Zechariah Chafee, an outstanding civil libertarian scholar, who in his *Three Human Rights in the Constitution* (1956) said: ''Foreign correspondents and lecturers on public affairs need first hand information. Scientists and scholars gain greatly from consultations with colleagues in other countries.'' [12] In their annual report of 1958, the directors of the Associated Press commented, ''A totalitarian government can release only the good news and hide its failures. In a free country, it is the duty of the government . . . and press, radio and television, to report the facts, good and bad.'' [13]

Before the United States became embroiled in the Vietnam war, the government had begun to bottle up many a news story as a matter of routine. Entering into the terrible age of the atomic bomb at the very end of World War II, the general public was told that government must maintain a monopoly on atomic science or our enemies would learn too much. The well-respected journalists Joseph and Stewart Alsop wrote, "Our people are ceasing to be the masters of their fate, because our government allows them to know less and less about the influences that control their fate. The American people cannot make the great national decisions any longer, because they do not get the essential facts." [14]

Hanging onto wartime habits, the government started to "classify" all sorts of military, industrial, and scientific information. Soon a web of restrictions with designations from "Top Secret" down to "For Official Use Only" or "Confidential" began to connect a complicated and sometimes unfathomable design of secrecy. A blue-ribbon Committee on Classified Information, appointed by Secretary of Defense Charles E. Wilson in 1956 to reform the system behind the secrets, could only point out the obvious problems of overclassification without suggesting a realistic remedy.

Congressman John E. Moss, indefatigable head of a special House of Representatives subcommittee looking into information blockages in the government, decided that not all was in order: "The information may be withheld by misusing the security stamp in order to avoid embarrassment or unpleasantness. In other instances, agencies apparently have discontinued information of a non-secret nature on the pretext of economy." He told his colleagues that he had reports of officials instructed not to associate socially or even talk with responsible writers and that one complaint, at least, charged that reporters who filed unfavorable stories were investigated and intimidated.[15]

Reporters as well as other citizens were routinely denied the right to travel to certain countries—the People's Republic of China and Castro's Cuba, for example—because the government said it was duty-bound to protect the personal safety of travelers carrying United States passports. There were other reasons, of course, the primary one being the official view in Washington that traffic with such "Red" governments was not good.

In 1961 the Committee on Government Operations issued this conclusion, supported by much documentation: "Secrecy is the handmaiden of bureaucracy, especially military bureaucracy. It has so pervasive an effect that all government becomes infected with the urge to restrict—even those routine agencies which should be wide open to the public." [16]

Reminders of past events which typically show how the battles for press freedom go are essential if we are to comprehend the basic point. It is this: There is much evidence to indicate that the battles are not being won, clearly and definitively, by the advocates of freedom. There is a creeping paralysis which seems to be more dominant in the body politic with each new generation. Unless we accept this realistic appraisal, we could all awaken one day to find freedom lost *here!*

THE SPIRIT BEHIND THE VIETNAM WAR AND WATERGATE

All too common today are news stories which feature facts or good guesses supporting the thesis that we have been lied to, misled, spied upon, or generally abused by this or that governmental agency. The Central Intelligence Agency, the Federal Bureau of Investigation, the Atomic Energy Commission, the Department of Defense, and the Executive Office of the President have become usual sources for investigative reporters looking for examples of usurpation of private privileges and public rights. After the Nixon administration, the dark cloud composed of "credibility gap" hot air, burst.

In the late 1950s reporters were complaining openly that government "security investigations" were ordered as reprisals against the representatives of the press who annoyed officialdom. Telephones were tapped, the reporters' contacts and friends were subjected to harassment, and "offending" reporters were cited by important bureaucrats as having earned the grave displeasure of the powers that were. In short, calculated steps were taken to stop the reporters from doing their job.[17]

With a mania for classifying documents overcoming normal respect for the public's right to know, government officials—notably those in the defense, intelligence, and foreign affairs communities—started to withhold information as a matter of routine. Beginning with the "Manhattan Project," which produced the first atomic bombs, a new concept emerged from bureaucrats with high public responsibility. We might call it the "illegitimate concept of security" and formulate it briefly this way: *That public is best which knows least about the most important matters of state.*

Long accustomed to the "handout" system, an endless chain of news releases issued through departmental information officers, reporters were not always quick to realize what they were missing. Periodic demands were made by press associations for declassification of documents or for quicker access to officials, but the majority of the reporters saw nothing pernicious in the overall system. Few members of the Fourth Estate believed that the government was getting into the habit of sponsoring secrecy for secrecy's sake and as a way of keeping official mistakes away from the prying eyes of newspeople.

Censorship, it was commonly assumed, was a necessary evil, not an administrative way of life. What had not been properly evaluated by politicians or by the press was the transformation of what can be called the governmental system. Defense spending, growing by leaps and bounds, gave the military a political lever over Congress and the economy that had never been contemplated. Technology, with its computers leading the paraphernalia of modernity, developed quantitatively while political knowledge seemed to stagnate. Government became the biggest spender, builder, instigator! Bureaucrats had to wear giant-sized boots, but there were no adequate checks on what they did with their normal-sized brains. The federal budget became the central factor of political life. During his Presidency, Harry Truman had to worry about a budget nearly 90 billion dollars huge. President Ford, after the debacles of Vietnam and the Middle East and the collapse of the world and domestic economy, worried about a budget nearly four times that size. In early 1975, the estimated annual *deficit* predicted for the year ahead was approximately 52 billion dollars.

Equally crucial to the change of the governmental system was the political habit of defining all matters in elemental electoral terms. Politicians, the press, and the general public went on naïvely believing that everything could be adjusted, no matter how serious the situation, if new politicians replaced old politicians.

In additon, there was a commonly held myth: that our nation, in any international crisis, was on the moral side; that all our leaders, nurtured by the roles of the United States in both world wars and in Korea, would strike the moral position in any future situation. It was popularly believed that by virtue of national history, philosophy, and political theology, this nation would automatically stand for the right, for the ultimate good.

We were in trouble at the end of Eisenhower's administration. He knew and warned us about one problem, in his farewell speech as President, when he pointed out the dangers posed by the military-industrial complex. The warning, seemingly so out of keeping with his easy-going approach, perplexed politicians and the press. Actually it was not all that peculiar. In the campaign of 1952, disturbed by Senator McCarthy, Eisenhower addressed a rally in Buffalo, New York, on the evils of "rabble-rousing." [18] They might have been startled as well, without really taking into account his message.

More than a decade later, George Ball also gave public warning unsuccessfully, emerging as the voice of dissent in the National Security Council as he tried to turn the Johnson inner circle away from further massive involvement in Vietnam in autumn of 1964. He was a dissident very lightly regarded by those who planned with an inner serenity that came from a confidence that they knew just what they were doing.[19] As it turned out, their administrative follies plunged the nation deeper into a morass in Southeast Asia.

To understand how the national administrations of this great republic came to tamper with press freedoms to the point that the political pendulum began to swing from democratic to autocratic, one must recognize that the leaders of the executive branch began to think in terms of power most of the time. Heady was the chemistry of power to those who understood the moral virtue of the United States as unquestionable, the physical power of the United States incomparable, the economic power of the United States indomitable. They judged social and philosophical attainments of the United States to be at the highest level attained by any civilization.

Add to the mixture an illogical disregard of valid and scientific evidence about the finite resources of this earth and the fantastically disproportionate consumption of the limited natural treasures in the United States. Such a chemistry of power produced the illusion that this government possessed the moral and political right, indeed the duty, to intervene around the globe whenever events in other countries took a turn deemed unacceptable to our President and the Cabinet advisers who directed national government agencies with seemingly endless power.

Contributing to the political illusion was a failure to take into account the near misses of our foreign policy, through which we were only fortuitously spared overwhelming disasters. Included in that category are our tragic errors in Indochina, when the French were attempting to stem the tides of nationalism; and the Korean police action, a series of diplomatic and military catastrophes superficially glossed over with publicity about a negotiated settlement with the enemy authorities. Regarding each of these protracted wanderings into the political and military swamps, it was the habit of officialdom to publicize each step as one taken on the road to *victory*. So long as public accountability was delayed, the government sought out new adventures, trying to make up losses like a habitual gambler throwing all on the next roll of the dice.

No change of policy or procedure would be forced on our leaders until the public saw through the false assumptions and the illusions. The media of mass communication would have to provide the investigative reporting necessary to the formation of educated public opinion. As the representatives of the press began to see the light, they found themselves increasingly under suspicion by the autocrats of the executive branch. As they reported more and more of the truth, they began to run up against the executive autocrats' cohorts in the legislature. Having been party to the events being reinterpreted, the powers that were of the Senate and the House of Representatives began to assume that they were in the circled wagon trains of our old west, under attack by the primitives of the press.

Truth is not a gentle teacher. Often truth is cruel to the extreme, shoving and pushing the reluctant to change. Sometimes truth is the enemy of the powerful who are so fearful of change that they take up arms against armies of facts. With few exceptions, the national leadership chose to fight a press that became more disillusioned with power at the same time that it grew more competent. The *increase* in press corps competency should not be taken as an inference that reporters became exceptionally adept at handling government leaders who viewed their sharper criticisms of policy as unpatriotic or even verging on the criminal. For a while, government excesses outpaced media changes for the better.

A tragic result of the domestic struggles was the unwitting support given to mindless radicalism by the government. Seeing all critics as enemies, it expended a great deal of official effort in snooping and sneaking. The reasonable and rational could find few allies. Ultimately, the truth did out, but at a terrible social cost—a divided society, a drawn-out bloodletting which spread from country to country in Southeast Asia, the collapse of the Western military alliance, decades of upheaval in the Middle East, destruction of the economic base upon which we and the rest of the technological world had come to rely, and a tragic loss of faith by our friends in other nations in the moral standing of the United States.

Any objective review of the literature on the Vietnam war and its aftermath reveals startling aspects of American political leadership in the 1960s and early 1970s. Despite the misjudgments and misunderstandings prevalent in the White House and Congress, Presidential dedication to the constitutional standards affording fair play to individuals and groups was accepted as an article of faith by the majority of the citizenry between 1960 and 1972. Presidents, like the Supreme Court, were assumed to be above political chicanery. Champions and detractors of Truman, Eisenhower, Kennedy, and

Johnson argued primarily in terms of differences over political goals and not of whether the chief executive was corrupted by power. We had to recognize our vulnerability to usurpations of the public authority.

The Watergate story and Nixon's leadership regarding Vietnam cannot be separated. One of the most pernicious elements of both consisted in the deliberate distortions peddled to the press by the administration. Another was the outrageous lying by the White House powers-that-were, in their attempt to outflank the many "enemies" who haunted their minds.

We had come a long way from the time of the U-2 incident, when President Eisenhower decided that no cover stories would be allowed after the capture of Francis Gary Powers by the Soviets. Taking full responsibility, Eisenhower had admitted that the United States government had lied in first denying any involvement in overflights of the Soviet Union. We had come a shorter distance from the Johnson administration, when the sincerity of the President did not discourage the military and their allies from fabricating events and conclusions as to their meaning, in order to sustain what dwindling popular support remained at home for the war efforts in Vietnam.

The major difference between the record of Nixon and his closest White House associates and the records of his predecessors is that he and his cronies manufactured distortions as products of administrative routine. The cancer which began with public illusions after World War II had spread not only to the underlings trying to cover up their failures but to the chief steward of the public trust at 1600 Pennsylvania Avenue.[20]

The step-by-step progression down the political road to the Nixon catastrophe was not handled well enough by the press. One plausible excuse is the weakness of the press corps on matters requiring expertise in political science, history, public administration, economics, business and corporate management, finance, and so on. If we consider reporters to be explorers after new fact, most of those of the past several decades have had to stop when their very ordinary road maps failed to help.

Historians like the eminent Arthur M. Schlesinger, Jr., can point out to today's students that Harry Truman exceeded his constitutional authority in committing American troops to the defense of South Korea in 1950, on the strength of a proposed United Nations resolution still in draft. He can observe that Eisenhower's extension of the concept of executive privilege to "a new and virtually unlimited category of information which he declared deniable at presidential will" was highly dangerous; and that the Tonkin Gulf resolution was rushed through Congress at Johnson's behest in August 1964 "in a stampede of misinformation and misconception, if not deliberate deception."[21] His examples of the "Imperial Presidency" and its growth make even more clear the press's failures to follow what was behind events. Earlier, the public's inability to *absorb* events was discussed. In a similar vein, the press has shown no great talent for getting away from facades and superficialities. So often the press corps has been content to deal with patterns of political events as chance *happenings*.

The general public learned more from the actual journals of the White House power brokers, the Pentagon Papers, and the Watergate Tapes than it had from decades of inadequate reports from the press.

One, and only one, of the many reasons for the press inadequacies noted is worthy of special mention here. After World War II, much of the power which flowed to Washington stayed there. That city became *the* center for national news. Becoming accustomed to dealing personally with the most powerful government officials in order to get their stories, reporters began to see themselves as somehow part of the power game. Many of the leaders of the journalistic profession became officials from the private sector dealing with officials from the public sector, and interests of the two groups drifted closer together and often merged. One result of the situation was that the press lacked sufficient interest in self-criticism.

Too many nationally known reporters failed to maintain the minimum necessary professional detachment from public officials whose activities they were covering. When they got exclusives on "inside" information, by benefit of personal friendships, there was always the temptation to exempt individuals from harsh publicity. The tendencies toward personalizing stories that should have been treated more objectively grew worse as the rewards of personal association led to more and more "scoops." No wonder, then, that some reporters ignored some stories, as did their bosses.

Lester Markel, whose professional experience includes forty years as Sunday editor of the *New York Times,* reports in his recent book *What You Don't Know Can Hurt You* (1973) on an eight-day survey he conducted of twenty morning and five afternoon daily newspapers, to determine how well they covered national and international news between June 9 and June 16, 1969. Markel concluded that in an important news week (Midway conference between Nixon and Thieu of South Vietnam, French national elections, the annual party congress in Moscow, campus disorders at home, congressional debate over the antiballistic missile system, increase in the discount rate announced by the Federal Reserve Board), these surveyed papers performed poorly. Regarding the stories on the Midway conference, the French election, and the discount rate, he rated three newspapers out of the total as worthy of a 50 percent rating for coverage; two he rated at 25 percent, and the twenty other papers "performed even more poorly." In Markel's opinion, the other three top stories "were barely covered, despite their importance." In July 1971, he rechecked the same newspapers and found that "the pattern apparently had not changed." [22]

PROTECTIONS FOR REPORTERS

Confidential information has turned out to present as many problems for news seekers as for news makers.

On June 29, 1972, the U.S. Supreme Court, in a 5 to 4 decision, ruled against reporters Earl Caldwell (*New York Times*), Paul Branzburg (*Louisville Courier-Journal*), and Paul Pappas (WTEV-TV, New Bedford, Massachusetts), denying their claim that they could refuse to divulge what they considered confidential source data to grand juries.

Caldwell had written stories about the Black Panthers and other black groups regarded as militant by popular press organs and general public opinion. When a federal grand jury subpoenaed the tape recordings and notes he has made during his interviews, he refused, as he put it, to protect credibility with sources. Before the Supreme Court ruled, the case had been subject to opposing decisions in the federal courts. The district court ruled that Caldwell had to answer the grand jury subpoena; the court of appeals found that he did possess a qualified privilege as a newsman and could refuse to appear.

Pappas had refused to answer the questions of a Bristol County, Massachusetts, grand jury about what he had seen in a Black Panther meeting place. The supreme judicial court of the state held for the grand jury. Branzburg, who had worked on a story about two young men who synthesized hashish from marijuana, had agreed to keep their identities confidential in order to get the interview and pictures.

Justice Byron R. White wrote the majority's views. He was joined by Chief Justice Warren E. Burger and Justices Harry A. Blackmun, Lewis F. Powell, and William H. Rehnquist. White declared that "the great weight of authority is that newsmen are not exempt from the normal duty of appearing before a grand jury and answering questions relevant to a criminal investigation." [23]

Champions of a liberal view of press freedom bridle at what they consider White's gratuitous words about the news-gathering profession on points not directly connected with the facts of the cases. They imply that political views bore heavily on the decisions. Particularly incensed at White's interpretations of "issues against the news media which were not even litigated" and his "statements of constitutional policy which, consciously or unconsciously, appear to misrepresent existing constitutional law to the detriment of the media," Fred P. Graham and Jack C. Landau urged journalists not only to carry on against censorship in the courts but to "seek a redress of their grievances at the legislative level." [24] Graham is a former *New York Times* reporter working the Supreme Court beat, and Landau is a Supreme Court reporter for the Newhouse Newspapers. As media specialists on legal procedures and issues, they have noted that other groups of professionals have privileges which allow them to avoid testifying before grand juries and to invoke protections of confidentiality. Cited are "more than 300,000 attorneys who may . . . invoke the attorney-privilege to protect confidential information from clients which might solve a case of heinous murder or treason. Also cited are special exemptions afforded physicians and clergymen.[25]

There are numerous state laws protecting confidentiality. As of 1972, nineteen states had "shield laws" giving reporters the privilege of refusing to identify news sources.

Commenting on the Caldwell case, the distinguished journalist Norman E. Isaacs (a past president of the American Society of Newspaper Editors) wrote that Justice White "used the judicial knife to slice the idea of a newsman's immunity privilege into bloody strips."

A few excerpts from White's opinion serve up the grounds for such anger:

> The administration of a constitutional newsman's privilege would present practical and conceptual difficulties of a high order. Sooner or later, it would be necessary to define those categories of newsmen who qualified for the privilege, a questionable procedure in light of the traditional doctrine that liberty of the press is the right of the lonely pamphleteer who uses carbon paper or a mimeograph just as much as of the large metropolitan publisher who uses the latest photocomposition methods. . . . Almost any author may quite accurately assert that he is contributing to the flow of information to the public, that he relies on confidential sources, and that these sources will be silenced if he is forced to make disclosures before a grand jury.

There were two dissents in the Caldwell case. One came from the pen of Justice William O. Douglas. The other three dissenters were Justices Potter Stewart, William J. Brennan, and Thurgood Marshall. Douglas, as direct as could be, countered White:

> The function of the press is to explore and investigate events, inform the people what is going on, and to expose the harmful as well as the good influences at work. There is no higher function performed under our constitutional regime. . . . A reporter is no better than his source

> of information. Unless he has a privilege to withhold the identity of his source, he will be the victim of governmental intrigue or aggression. If he can be summoned to testify in secret before a grand jury, his sources will dry up and the attempted exposure, the effort to enlighten the public will be ended.[26]

Will the truth out? The Caldwell and related cases indicate much trouble ahead for independent media. There are multitudes of examples showing that repression of the press is becoming a habit in these times of crisis.

Even so-called victories are not clear. Despite all the fuss made to get the Pentagon Papers into print—the long legal struggles waged by the *New York Times* and the *Washington Post,* which culminated in the public's learning how policy was made at the highest level of our government—a clear conclusion cannot be drawn about the status of press freedom.

Regarding the Nixon administration's allegations in support of executive privilege, in the context of the enormous effort to prevent publication of the bulk of the forty-seven volumes, the Supreme Court superficially favored the press. By votes of 6 to 3 in *New York Times v. United States* and *United States v. Washington Post* (1971), the Court noted that the government had not provided the heavy burden of proof necessary to support prior restraint.

In basic terms, the Pentagon Papers affair was only a skirmish in a war of great magnitude. The administration lost because it had stupidly dared to attempt to deny *all* to those who sought the truth. The affair proves nothing conclusive about selective censorship or about the chances of success for more talented *but* equally secretive and monopolistic administrators.[27]

Looking back on some of the big stories of the past decade which had to be torn from behind the cloaking veils of the cover-up artists, one does not turn suddenly optimistic. We all recall the My Lai massacre in South Vietnam and the dogged attempts made by the military to keep the horrible truth a secret from the public.

Usually, those who demand confidential material about sources stress only their constructive concerns to end the abuses related to the public through the media. Nevertheless, if reporters destroy their *ability to get confidential material* by telling all to grand juries and legislative committees and police officers, there will be even more sordid stories that will never be brought to the attention of reformers and defenders of the law. Each case is different, and each demand for information must be examined on its own merits. Should Joseph Weiler, reporter for the *Memphis Commercial Appeal,* have given in to a subpoena of a state legislative committee by revealing the identity of the sources behind his series on brutality against mentally retarded children in a state hospital? When he refused, he was cited for contempt! Should Samuel L. Popkin, the Harvard professor, have agreed to the order of a federal grand jury to give the names of persons "with whom he discussed the Pentagon Papers"? He went to jail for eight days. Should Stewart Dan and Ronald Barnes, respectively reporter and cameraman employed by WGR-TV in Buffalo, New York, have obeyed a county grand jury's order to tell what they witnessed when they were inside Attica Prison during the inmate rebellion of 1971? [28]

Will the truth out, despite repressive tactics of intimidation, such as undertaken by the Nixon administration against Daniel Schorr, the highly reputed veteran correspondent for CBS News? On Friday, August 20, 1971, an agent of the Federal Bureau of Investigation showed up at CBS News headquarters in Washington, to investigate Schorr. The reason given to Schorr and to the bureau chief, William Small, was that he was being considered for "a high government position." When asked, the FBI man could not even give a clue as to what the job was. The visit was part of a "full field investigation." Employers (past and present), neighbors, colleagues, and others were questioned. Actually, the business about a government position was most probably pure fiction, an invention useful to those in the Nixon inner circle who disliked Schorr's reporting. It was harassment, direct and chilling to all advocates of personal rights as protected by the Constitution. In 1972, Schorr wrote, "I am constantly asked whether my 'FBI shadow' is with me, whether it is safe to talk with me on the telephone, whether I am still 'in trouble with the FBI.' " [29]

Will the truth out, *because* the Nixon administration's efforts to politicize broadcasting (through harassments intended to curb or eliminate reporting that the White House found distasteful or obnoxious) taught us all a hard lesson about capricious uses of power?

Clay T. Whitehead, director of the Office of Telecommunications Policy during Nixon's tenure, scolded public television broadcasters in an October 1971 speech to the National Association of Educational Broadcasters. He criticized their allowing the Ford Foundation to "buy $8 million worth of [public affairs] programming on your stations." Later, appearing on the "Today" television program, he asked if "public broadcasting [should] be doing the same kind of news coverage . . . that we get in a pretty good way from private television." A year afterwards, he attacked network news, talking about "ideological plugola" and "so-called professionals who confuse sensationalism with sense and who dispense elitist gossip in the guise of news analysis."

Whitehead, of course, is entitled to his opinions. But in his official role, what he said about news and commentary on the networks caused many a journalist, network administrator, and concerned citizen to wonder if the federal government's licensing procedures and powers were not being waved as a warning. It seemed quite clear that his comments about broadcasting responsibilities were part of a blanket of influences that he was weaving for the administration.[30]

Walter Cronkite, the doyen of anchormen at CBS News, appeared before the Subcommittee on Constitutional Rights, U.S. Senate (former Senator Sam J. Ervin, Jr., Chairman), on September 30, 1971. He summed up the problem:

> There are the pressures exerted by high government officials who suggest that if we don't put our own house in order, that is, report the news the way they would like it reported, then "perhaps it is time that the networks were made more responsive to the views of the Nation, and more responsive to the people they serve."
>
> The speaker might, indeed, disclaim any intention of censoring broadcast journalists, but when the speaker is a high official of the administration that appoints the [FCC] Commission that holds life or death power over the broadcast industry, a broadcast journalist and his employer might be excused for thinking that it sounds like a threat.[31]

FREEDOM OF INFORMATION LEGISLATION

1966

Ramsey Clark, Attorney General of the United States in 1966, issued a memorandum to help federal executive branch officials understand the Public Information Section of the Administrative Procedure Act, signed into law by President Johnson on July 4. Clark advised that the legislation "leaves no doubt that disclosure is a transcendent goal, yielding only to such compelling considerations as those provided for in the exemptions." The new law, he said, indicated congressional and Presidential "key concerns," including:

> . . . that disclosure be the general rule, not the exception; that all individuals have equal rights of access; that the burden be on the Government to justify the withholding of a document, not the person who requests it; that individuals improperly denied access to documents have a right to seek injunctive relief in the courts; that there be a change in Government policy and attitude.[32]

At the time of its passage, the law was particularly opposed by the federal intelligence and law enforcement officialdoms. The military, worrying about release of information bearing upon negotiation and administration of defense contracts with suppliers of equipment, were not sympathetic. The national security "blanket" did not cover enough when a snooping press inquired into embarrassing "cost overruns" or into problems behind the manufacture of such major items as the F-111 fighter-bombers or the huge C-5A transports. The Navy disliked the attention paid to its tremendous investments in a submarine torpedo that failed to live up to the marvelous descriptions of potential effectiveness circulated whenever more money was needed.

The military were not alone in their suspicions about the new law. The greater majority of the leaders of federal departments were not favorable. Since 1966 the major proponents of the legislation, consumer groups, and the media have had to share in the disillusionment because so little positive action ensued. The burden of proof, it turned out, was on the press and not the government. It is estimated that only a half dozen major attempts to get key information under the terms of the 1966 act were undertaken by the press. Carl Stern of the NBC-TV staff kept patiently plodding on with litigation, for some twenty months, to get a key story about how the Federal Bureau of Investigation used a counterintelligence group to infiltrate New Left organizations.

Congress was as frustrated as the press corps in trying to ferret out information from the executive agencies. Senator Howard Baker, Jr., of Tennessee, the Republican deputy chairman of the Senate's special committee investigating the Watergate case, has complained that he and the other committee members were unable to get adequate help from the Central Intelligence Agency when it came to declassification of information sought.[33]

Testifying before the subcommittee of the Senate's Committee on the Judiciary looking into government information practices, John H. F. Shattuck, staff counsel of the American Civil Liberties Union, argued against excessive use of the exemptions in the 1966 legislation, through restrictive interpretations imposed by the administrators and by the Justice Department leaders, not excluding Ramsey Clark. He reviewed the "obstructive administrative procedures for processing requests for information" which had grown since 1966 and to the spring of 1973. "These include complicated agency request forms, exorbitant filing and reproduction fees, an unreasonable degree of specificity in identifying requested documents, refusals to separate non-exempt from exempt information, and unconscionable delays in processing initial requests and administrative appeals."[34]

Harold E. Hughes, then U.S. Senator from Iowa, told the investigating legislators (May 8, 1973) of typical frustrations he had endured as a member of the Armed Services Committee of the (*greatest deliberative body in the world*) Senate. He had derided the "illegal bombing" authorized by General Lavelle in Southeast Asia, under the cloak of the "massive cover-up of information" engineered by high officials supposedly controlled by the Pentagon's high brass; only persistent efforts by the Congress had forced the Pentagon to reveal the full story. Hughes told of his repeated requests to the Defense Department "for monthly American air combat sortie and bomb tonnage figures for each of the several countries of Indochina." He had received the information, albeit incomplete, but it was classified as secret. "Why," he demanded to know, "should this information be withheld from the American people? The people bombed know what has happened to them; Hanoi radio reported sortie figures. The only ones who did not know whether or how much we have been bombing in Laos or Cambodia or North Vietnam were the people of the United States, who were being asked to pay for and support this military strategy." [35]

The principal author of the 1966 Freedom of Information law looked back in 1972 at the many developments which did not bode well for the public. Congressman John E. Moss cited very restrictive judicial interpretations favoring the secretive in government. Against that trend, he argued that "Congress intended *all* government documents to stand up to the test of careful evaluation and review, and by an independent reviewer at that." [36]

1974

On October 17, 1974, President Gerald R. Ford vetoed a new Freedom of Information Act which, by virtue of amendments to the 1966 legislation, allowed for greater public access, by right, to government information. The sympathies of the House and the Senate were clearly not in accord with his action. On November 20, two-thirds of those present in the House voted to override (371 yeas to 31 nays). The next day, the senators voted 65 to 27 to affirm the new law.

Despite the specific exemptions provided, President Ford had objected. The exemptions deal with secret national security or foreign policy information, internal personnel practices, information specifically exempted by law, trade secrets or other confidential commercial or financial information, inter- or intra-agency memos, personal information, personnel or medical files, law enforcement investigatory information, information related to reports on financial institutions, and geological and geophysical information.

The new law permits court review of national security classifications. In addition, the courts are granted discretionary authority to award court costs and attorneys' fees to successful practitioners. A procedure is established for disciplinary action when a court finds that a federal official has acted capriciously or arbitrarily in withholding information. A time limit of ten working days is set for an agency to respond to a request for information; twenty days to answer an appeal from an initial request; thirty days to respond to a complaint filed in court under the act.

Government agencies are required to maintain an index of documents so that citizens know where and what to look for. Excessive charges are forbidden; agencies may charge only what it costs to provide the requested information. The Civil Service Commission is required to start proceedings to determine if disciplinary action is warranted in cases where a court holds that an official acted "arbitrarily or capriciously" in denying information.

In his veto message, Ford went over old ground. Despite its provisions, he remained "concerned that our military or intelligence secrets and diplomatic relations could be adversely affected by this bill." He objected to any judicial role in the classification processes. He worried about the confidentiality of FBI files and thought that the ten- and twenty-day requirements imposed on agencies were unrealistic "in some cases." [37]

The central issue now is not whether the law is improved. The public and its important legislative, judicial, and mass media representatives are anxious to know whether, after all the terrors and traumas of recent years, officialdom from the President down will obey the law in spirit and therefore in letter.

ACCESS RIGHTS VERSUS CONSPIRATORIAL POLITICS

Fairness or Foul Play?

After all this country has been through in recent decades, it is not surprising that versions of the "conspiracy theory" are so often presented by analysts rationalizing backgrounds to particularly difficult and dramatic events. Allegations that known outcomes resulted from deliberate planning, usually by malevolently motivated groups, satisfy deep psychological needs of broad publics who crave direct and uncomplicated explanations.

The conspiratorial rationalization makes it easier for the explainer to adjust all known information so that blame for an outcome is definite. Old interpretations and previous sources of information are downgraded. The new information is offered in such a way as to cancel out explicit values which had been drawn from evidence. The manner in which old and new information is combined results in the creation of a wholly new primary message.

When individuals try to figure out what happened in any given situation which makes them worried about personal security, they try to connect their assumptions about the world to new problems. Social science researchers have developed models dealing with ways in which individuals orient themselves to messages when trying to make sense out of complex environments.

Professor Robert Axelrod, of the University of California at Berkeley, a foremost interpreter of the processes of perception and cognition, in interpersonal and interpublic communication, has recently endeavored to apply a "schema theory" of idea transfers to politics. Informally, he defines a *schema* as a "pre-existing assumption about the way the world is organized." Each of us tries to fit new information into the schema we rely upon. "If the new information does not fit very well, something has to give." More precisely, Axelrod defines a schema "in terms of the set of all specifications which have certain stipulated properties."

It is important that we keep in mind certain of his admonitions relating individual belief systems to perceptions of new information and to ways in which decisions are reached. Discussing a "rather common problem," he concludes that "the schemata people use are often larger than they think, and therefore less easy to disconfirm." Also, "historical analogies seem to be drawn by selecting the single previous case which provides the satisficing fit to the present case." [38]

All these words of caution are reasonable because of recent demands for the abandonment (because of abuses) of the Fairness Doctrine, as now applied to radio and television but not to the print press. Keeping in mind the benefits, we must examine proposed changes carefully to ascertain if there is a good case for reform in regard to any situation brought to our attention. *We must be scrupulous, as well, not to shy away from the conspiracies that are uncovered.* With that in mind, the background of the landmark U.S. Supreme Court case *Red Lion Broadcasting Co., Inc., v. Federal Communications Commission; United States v. Radio Television News Directors Association* (1969), is very interesting.

The basic outline of the case, as known to the federal judiciary and the public at the time of the decision, is not particularly complex. On November 25, 1964, radio station WGCB of Red Lion, Pennsylvania, played a tape produced by the Christian Crusade organization in Tulsa, Oklahoma. Included in that broadcast was an attack by the Reverend Billy James Hargis, a well-known right-wing radio preacher, upon Fred J. Cook, the prolific writer. Hargis, an enthusiastic Goldwater supporter in the 1964 campaign, believing him to be the leader needed to assure "the survival of a free America," was angered and agitated by Cook's anti-Goldwater book *Barry Goldwater: Extremist of the Right.* He was enraged, too, by the article "Hate Clubs of the Air," published in *The Nation* magazine, in which Cook had "classified Hargis as a bigot." During the controversial taped broadcast, Hargis labeled Cook "a professional mudslinger" who was dishonest, had falsified stories, and had defended Alger Hiss.

One of the most respected commentators on mass media developments, Fred W. Friendly (now Edward R. Murrow Professor of Journalism at Columbia University), offers new and persuasive information and arguments, based upon his extensive investigative reporting of the *Red Lion* situation. Cook, he alleges, was part of a White House-directed effort to suppress extremist "right-wing" radio commentators. He had mailed "200 letters; about 50 of the stations agreed to air a reply." WGCB sent Cook the station's rate card and the note, "Our rate card is enclosed. Your prompt reply will enable us to arrange for the time you may wish to purchase."

After WGCB's decision to refuse free time, the FCC ordered the station to supply such time for Cook's reply. The station management refused, appealed the order in the U.S. Circuit Court of Appeals in Washington, and lost its case. Following that decision, the FCC issued new and more stringent regulations regarding personal attacks and political editorials about candidates for public office. One of the new rules stipulated that, within one week, the stations had to notify persons who were the subjects of attack and had to offer them free time to reply. A possible penalty of $1000 was set for failure to provide such notification.

At about that time the Radio Television News Directors Association decided to fight the Fairness Doctrine, in opposition to the new personal-attack regulations of the FCC. The organization's case was heard by the Seventh Circuit Court of Appeals in Chicago, which ruled against the FCC's new rules.

With one circuit court of appeals finding for Mr. Cook in his legal battle with WGCB and another ruling against the FCC regulations, the Supreme Court decided to consolidate the two disputes, when it accepted these Fairness Doctrine issues for review. The access aspects of the cases were of special interest to the highest court's judges, as were the free speech and free press aspects. The FCC was upheld, and the Seventh Circuit Court rulings against the new administrative regulations were reversed.

Fred W. Friendly contends that there is evidence that the White House during the Kennedy Administration monitored broadcasts by so-called right-wing commentators in order to prompt requests to radio stations by people who would demand the right to reply under the Fairness Doctrine. According to Friendly, several of the President's "strategists" were concerned that the "nuclear test-ban treaty with the Soviet Union was being jeopardized by right-wing commentators who denounced the treaty and argued against its ratification." Friendly also asserts that in October 1963 Kenneth O'Donnell, one of President Kennedy's key aides, set in motion a monitoring program of the "radio right," the plan being to "harass the radio stations by getting officials and organizations that had been attacked by extremist radio commentators to request reply time, citing the fairness doctrine."

Professor Friendly ties together programs and interests of the White House, the public relations firm of Ruder and Finn, the Democratic National Committee, and the tax-exempt Public Affairs Institute, all of which, he alleges, were in a coalition to reduce the efficacy of right-wing broadcasting. The coalition was informal but effective, we are led to believe. For example, Ruder and Finn organized a group which Friendly describes as "a bi-partisan front organization called the National Council for Civic Responsibility." Later it became the National Committee for Civic Responsibility of the Public Affairs Institute. Friendly says that the Public Affairs Institute was moribund when the work of radio station monitoring began but, as a tax-exempt organization, was seen to be of potential use to the Democratic National Committee. According to Friendly, the head of the Institute was told by a leading Washington lawyer close to the White House since the Roosevelt administration, and by the chairman of the Democratic National Committee, "We got the money and you got the tax exemption and we need you to fight those right-wing radio extremists."

Fred Cook was hired to do research for Ruder and Finn, which was producing programs for the Committee for Civic Responsibility. His book about Barry Goldwater was subsidized by the Democratic National Committee, which "offered in advance to buy 50,000 copies." [39]

Friendly's charges of Kennedy-era Watergate-type manipulations ought to be studied in full by every concerned citizen.

If we accept his documentation and its implications, with all the sincerity with which it is offered, do we go one step further and say that, because of the intimidations of the press by the Kennedy, Johnson, and Nixon administrations, we ought to dispense with the Fairness Doctrine as it now applies to radio and television? Should we consider easing the present FCC rulings, moving away from broadcast-by-broadcast or issue-by-issue review, to regulation which deals with overall patterns of station fairness in the radio and television areas of mass communication?

No doubt the broadcasters would like that. On the other hand, there is virtue in the comments of the Reverend Everett C. Parker, director of the Office of Communication of the United Church of Christ: "The fact that people who are propagandists misused it doesn't make the law any less necessary to offset the monopoly broadcasters have on political and social comment. It is still a fair and proper way to handle the right of access to the broadcast station."[40]

Intimidation of the press, and attempted intimidation, is all too easily documented. Examples abound indicating governmental, business, and interest-group intrusions which go beyond appropriate democratic advocacy.

Representative Benjamin S. Rosenthal (Democrat, New York) asked the Justice Department to look into the Shell Oil Company's quest for a list of all radio stations which had carried the series of commercials prepared by Public Media Center (a nonprofit advertising company). Those spots were critical of the oil industry and favored the environmentalists' positions. One radio spot, entitled "Highway Robbery," expressed the view that the oil industry was earning excess profits. By September 1974 the Center's director estimated that some 700 stations had requested the material.

Shell was determined to obtain the list and even went through an intermediary to get one. There is no evidence that the company had knowledge that misrepresentation was a factor in the Center's decision to release the list to a telephone caller alleging that he worked for a unit of National Public Radio.

At that point the affair becomes cloudy. What was the Shell Oil Company purpose, vis-à-vis the radio stations involved? We know that Shell's vice president of commercial marketing entered the list as evidence when he appeared before the Senate Commerce Committee's Subcommittee on the Environment. The senators were considering data and views as to whether oil companies could take tax deductions for advertising dealing with "political advocacy" rather than simple product promotion. In 1973, Shell's advertising budget was 15.3 million dollars, of which 2.3 million dollars funded "so-called institutional ads" which could be taxed if the law is changed.[41]

Sensitivities about media coverage of news and views are certainly understandable. The oil companies of the United States have had to deal with unprecedented public pressures since the start of the current oil crisis in the autumn of 1973.

In an advertisement entitled "Musings of an Oil Person . . . ," appearing in the *Boston Evening Globe* of July 22, 1974, the Mobil Oil Corporation presented its views about some of its critics. Included was this commentary about the press:

> . . . Can't help wondering about the news media. Most try to be fair. But a few reporters, unencumbered by knowledge of our complex business, decline to let the facts get in the way of this story. Prefer to rely on such tired labels as "oil barons," "big oil," and powerful oil lobby. . . . Maybe some day they'll have second thoughts about their brand of "robust journalism" which in other times would never have gotten past a good copy desk. . . .[42]

HOW FAIR CAN MEDIA BE?

Generalities do not suffice when one talks about media fairness. Cases count more! When challenged for its 1972 documentary "Pensions," aired on its television network, the National Broadcasting Company defended the highly critical hour-long production. The FCC, responding to arguments that the program was biased, ordered NBC to present other material favoring the pension situations of the day. NBC successfully appealed the FCC ruling in the courts. Its arguments make good sense. The network view was that investigative journalism was jeopardized by the FCC demand. On the program it was made clear that good pension plans existed but, on the whole, workers were not adequately protected. It was nonsense for the FCC, an agency of the government, to take the view that a balanced picture of merits and defects of pension plans had to be presented. If the FCC order had been sustained, documentaries on crime and corruption would also have to be balanced, showing the splendid achievements of law enforcement agencies. It would be as fair to inform a robbery victim that seminaries are enjoying increased enrollments, when he wants to know if anyone is chasing the thug he had to face.[43]

Fairness to Whole Communities: The WNET-TV (Channel 13) Situation

On February 1, 1975, the FCC voted to conduct an inquiry into ways of providing more adequate television service for New Jersey citizens. Highly populated New Jersey has no major commercial television station, and so the populace has to rely upon New York or Philadelphia stations. There is one major educational channel assigned to the "Garden State," but its studios are in New York City.

New Jersey officials complain of inadequate coverage of state news as well as insufficient attention to the state's culture, economics, and politics. That is a very serious complaint when more than 7 million residents are involved. A petition to the FCC was initiated by the New Jersey Coalition for Fair Broadcasting, which consists of eighteen civic groups.[44]

On June 4, 1974, a petition was filed with FCC by the Puerto Rican Media Action and Educational Council, Inc. (PRMAEC), alleging discrimination by WNET-TV against Puerto Ricans and other "Latinos, Hispanics, and/or other Spanish-origin persons." PRMAEC claimed that about one million such persons dwelled in the viewing area of the station. FCC researchers decided that 5.9 percent of the New York City Standard Metropolitan Statistical Area population consisted of persons of Spanish heritage and that 2 percent of the New Jersey population could be so classified.

The curious aspect of the PRMAEC argument was insistence that "any programming which does not take into consideration the richness and uniqueness of the Puerto Rican culture will only be tangentially relevant to the needs, interests, and views of that community." Also, the group declared that because "Hispanics are predominantly Spanish-speaking, English-only programming is both irresponsible and incomprehensible to much of the Hispanic community.'' The FCC did not accept either contention. It noted that ''it is likely that many of those persons speak or have received training in English.''

PRMAEC listed the following programs as examples of what it found discriminatory in the educational station's offerings: ''Bill Moyers' Journal,'' ''An American Family,'' ''Who's Afraid of Opera?,'' ''Report to the Nation,'' ''The Great American Dream Machine,'' ''This Week,'' ''Opera Theatre,'' ''Vibrations,'' ''NET Playhouse,'' ''Masquerade,'' ''NET Journal,'' ''The World We Live In,'' ''What's New,'' ''News in Perspective,'' ''International Magazine,'' ''The President's Men,'' ''Spectrum,'' ''Cities of the World,'' ''Great Decisions,'' ''State of the Union,'' ''At Issue,'' ''Date Line,'' ''America's Crisis,'' ''Men of Our Time,'' ''Intertel,'' and ''Power of the Dollar.''

Anyone conversant with American television will recognize many of the cited programs as exceptionally important educational achievements in a medium dominated by commercial mediocrity. What, then, was behind the PRMAEC petition? It is clear that programming specially directed to Hispanic culture, needs, and interests was demanded, to replace much of interest to the general community consisting of all the varied groups which live together in the area covered by the WNET-TV signals.

WNET-TV responded that it provides programs of particular interest to Hispanic audiences. Listed as examples were ''Carrascolendas'' (a children's program); ''Realidades''; and Spanish-language versions of ''The Killers'' (a health series), ''Consumer Help,'' and ''The 51st State.''

To buttress its other charges, PRMAEC cited WNET-TV as discriminatory in its employment practices. The FCC decided that the organization had not proved that the station had so discriminated.

Against his fellow commissioners finding for WNET, Commissioner Benjamin L. Hooks dissented, lambasting the station's ''glaring deficiencies.'' Hooks's views are, on the whole, novel, in that they appear to add up to a rejection of the current approach of educational television as unfair. He says that "the time has come for a showdown with public television." Excerpts from his dissent follow:

> WNET's sin, one of arrogance, is to have concentrated its efforts on one minority group, the cultured, white cosmopolites, and too often neglected the enlightenment of other less fortunate minorities which it has a fundamental duty to serve.
>
> . . . Although I recognize that minority groups do derive great benefit from programs of mass appeal . . . and that general public affairs shows cannot always be broken down into ethnic points of view . . . there exists a simultaneous obligation not to elevate the general to the unreasonable impoverishment of significant minority groups.
>
> . . . By styling itself, preponderantly, as an electronic Harvard liberal arts course, public broadcasting has forsaken those less privileged and influential whose cultural and educational needs are far more on a "street academy" or community college scale. By aspiring to titillate the sensibilities and sensitivities of the twentieth century Renaissance man, it has overlooked the intellectual needs and sensitivities of that core of the population which, after years of third-rate education and cultural oppression, is just emerging from the chains of the eighteenth and nineteenth centuries. By disproportionately featuring the refinements of Western European vintage, it has slighted those whose heritage derives from Africa, Latin America and the Orient.[45]

Any critic of media fairness and its political ramifications has to take into account the angry views respecting the frustrations Hooks expresses and the media's failure to do a good enough job for minority groups. It would be incorrect to say that Hooks is opposed to cultural programming or wrong in his views of minority needs. Beyond the particular of *In re Complaint of Puerto Rican Media Action and Educational Council, Inc., et al. Against Educational Broadcasting Corporation, Licensee of Station WNET-TV*" lies a substantive reinterpretation of what fairness should mean. Commissioner Hooks's assault on the bias he considers implicit in the best of educational television presentations could, at the least, divert some funding and talent for new programs, fairer to important segments of the audience. Such a diversion would make good political sense as well.

Access and Citizens' Right of Reply

The U.S. Supreme Court, in a most significant recent decision, emphasized the special privileges and responsibility of the print press. In *Miami Herald Publishing Company v. Tornillo* (1974) the Court unanimously struck down a Florida law of 1913. Under its provisions newspapers in the state were required to print replies from political candidates attacked in their pages. Chief Justice Burger wrote for the Court, "The clear implication has been that any such compulsion to publish that which '*reason* tells them should not be published' is unconstitutional. A responsible press is an undoubtedly desirable goal, but press responsibility is not mandated by the Constitution and like many other virtues it cannot be legislated." The Florida law, he observed, failed "to clear the barriers of the First Amendment because of its intrusion into the function of editors."

The newspaper immediately involved, the *Miami Herald,* was defending itself against the complaint of a candidate for the Florida House of Representatives, one Pat L. Tornillo, Jr., who was also executive director of the Classroom Teachers Association of Dade County (a 12,000-member teachers' union). In an editorial very strongly urging voters' rejection of Tornillo, the *Herald* had called him "Czar Tornillo." About eight days after the editorial appeared, he demanded that the newspaper print a letter of his own in reply, or be in violation of Florida Statute 104.38 of 1913, known as the right-of-reply law. That law, in the criminal code, required newspapers to print, free of charge, a political candidates's reply to criticism of him which was published in the press. Additionally, the reply had to be accorded the same prominence as the initial criticism.

By its action, the Supreme Court voided a serious threat to press freedom. With so little precision found in law or common sense as to what constitutes an attack on an individual, the law, taken literally, would invite wholesale domination of the press by persons claiming a right to reply. Access rights in theory and practice would be mocked with the possibility of a thoroughly intimidated and controlled press.

Justice Byron White, in his concurring opinion, observed, "The First Amendment erects a virtually insurmountable barrier between government and the media so far as government tampering, in advance of publication, with news and editorial content is concerned.[46]

By Douglas E. Jamieson. © 1974 by The New York Times Company. Reprinted by permission.

MEDIA AND ELECTION TRENDS

5

Politicians are obliged to represent constituencies; to provide leadership on issues; manage the compromises and concessions necessary to ensure progress; seek equitable solutions to problems which produce difficulties for all involved; uphold and enhance ethical and moral principles; and administer public business in accordance with constitutional, legislative, and juridical standards. All this is expected of any politician subscribing to the democratic goals and republican structure of the United States.

Admittedly, this list of shibboleths would surprise large segments of the public whose members do not assign so much importance to the foregoing responsibilities of politicians. It is quite common for citizens to emphasize the publicity aspects of political life, with special attention to election processes. On that basis, politicians are obliged to pay attention to the demands of important pressure groups, be personable, understand and practice good public showmanship and the allied art of publicity, defend personal and civic virtue and be publicly known as a practitioner of such virtue, and win elections.

The contrast between the ''great principles'' and the widely held ''street corner'' interpretations of the political role alarms all serious observers of the political scene, despite the fact that there is no novelty in current misunderstandings. The general public in ancient Athens and in Imperial Rome harbored quite similar views.

Popular imagery of politics and politicians is reinforced by the mass media, especially the electronic. With so much television and radio time devoted to entertainment, public interests have been built up, all out of proportion to political realities, in politics as exciting electoral races. The majority of politicians today grew up in a television environment. Most do not even question the background to the preoccupation with vote solicitation. Culturally attuned to the situation, they concentrate on publicity campaigns for much of the time they are in public office, and for most of the time they are seeking to unseat and replace officeholders.

An unhappy side effect of the emphasis on spectacle is greatly increased public apathy and cynicism about public affairs. In an important manifestation, decreased percentages of eligible voters have gone to the polls in each of the recent presidential election years. For example, after the media-dominated presidential election campaign of 1972, the turnout of voters on election day had dropped by 6.2 percent as contrasted to 1968. Since the national campaign of 1960, the actual turnout at the voting booths had decreased by 8.7 percent.[1]

If we except the tremendous outpouring of emotion, rhetoric, and demands associated with the agitation over the American involvement in Vietnam, popular enthusiasm for most causes has waned during the past fifteen years. Even the fundamental fervor for civil rights has substantially moderated. And the change is partially understandable as we look back on the harrowing Vietnam conflict and the great domestic divisions that grew out of it (at the Watergate scandals with all the attendant manifestations of abuse of power, and at the major social disturbances that boiled up all over the land).

Typical adult citizens have come to their apathy out of fear and anxiety. They worry that money is the base ingredient of politics. They sense that the two-party system is endangered by the apparent fragmentation. They have become so accustomed to their role as viewers of politics via the mass media that they have lost any sense of participation. On every hand, they find evidence that politicians have been and are off on publicity binges in which many complicated matters are reduced to superficialities. They are told that political campaigning is a game reserved for the rich and that individuals of modest means ordinarily cannot run for public office unless backed by people with money. A great deal of the money expended in election campaigns is gobbled up by costs of mass media exposure of candidates, and television time is usually the most expensive of all.

The dispirited average citizens were probably at their lowest ebb as they watched the daily news breaks during the two or three months preceding Richard Nixon's resignation on August 9, 1974. By the last stages of the former President's stewardship, it had become abundantly clear that corruption had been rampant. Unethical and criminal actions, brought to public attention by the media, ranged from the cover-up of the Watergate scandal to outright lies about sources and uses of large campaign donations to distortions of what was actually on the famous tapes which recorded Presidential conversations.

With all that, there was also the dramatic radio and television coverage of the open hearings and meetings of the House of Representatives Judiciary Committee, which in late July 1974 voted three articles of impeachment against the President. The long public sessions of that committee were an appropriate sequel to the long and even more dramatic hearings of the special Senate committee which, beginning on May 17, 1973, worked to unravel the tangled web woven by high-level conspirators, their dupes, and their willing followers.

When the House Judiciary Committee was well along in the work culminating in the impeachment votes, the majority leader of the Senate, Mike Mansfield (Democratic, Montana), gave thought to how the Senate should fulfill its constitutional obligations vis-à-vis use of the public media for open sessions. Sensitive as he is to the proprieties requisite to the business of the Senate, he came out squarely in favor of television coverage:

We should recognize that if an impeachment does occur again, future generations would assess us by the propriety and fairness with which the Congress proceeded to reach a decision.

That is what gives the gathering storm immediate and historic significance. Whatever might be the final verdict is a secondary consideration. Most important would be the confidence reposed by the people in the integrity of the process. Either way, the result must be perceived as having been justified by the evidence and argumentation, otherwise the basic constitutional structure of the nation will suffer still another blow. . . .

There is no other means than television to convey directly to the people of the nation the gravity of these proceedings and the solemnity of the atmosphere in which they must be conducted. There is no other way other than by television to permit the formulation of a fully acceptable public judgment of the integrity by which this constitutional responsibility is discharged.[2]

REFORMING THE SYSTEM: TELEVISION'S POWERFUL PERSONALIZING IMPACT

As it turned out, the question of a televised impeachment trial became moot when Richard Nixon resigned. Nevertheless, by August 1974 the media had done much to assist in the processes of justice. Beginning with the first televised hearings of the Senate Select Committee on Presidential Campaign Activities (mandated to investigate, by a Senate vote of 77 to 0, on February 7, 1973), the sordid aspects of the Nixon administration were disclosed to the American people and to a world audience. (The author, visiting in London in the summer of 1973, remembers the widespread interest of English people, across the social strata, in the live television broadcasts which came to them late at night and in the early hours of the morning. Bleary-eyed from lack of sleep, his friends would recount the latest developments, expressing amazement at the willingness of American politicians to bare the nation's soul to save its legal substance.)

The chairman, Senator Sam J. Ervin, Jr., captivated audiences with his rustic charm, a veneer that did not diminish his great authority as a concerned constitutionalist. Indeed, many of his "downcountry" stories served to heighten attention on the paramount issues before the nation. All the members of the committee (Vice Chairman Howard Baker, Jr., Herman E. Talmadge, Daniel K. Inouye, Joseph M. Montoya, Edward J. Gurney, Lowell P. Weicker) became superprominent to the enormous television audiences as they commented on the issues or questioned witnesses. The legal staff, headed by Chief Counsel Samuel Dash, with Fred D. Thompson as Minority Counsel, was equally in the public eye, questioning witnesses slowly but surely to bring forth as testimony the details covered earlier in closed sessions.

Day by day, during long weeks of open hearings, the stories behind the Watergate break-in, the Houston Plan, the "enemies list," the White House Plumbers, Project Sandwedge, Gemstone, the Committee for the Reelection of the President (CREEP), the Milk Fund, and so on, were made very, very clear.

One by one the characters involved in the events paraded to the witness chair. John Dean, John Mitchell, E. Howard Hunt, Charles Colson, Gordon Strachan, Maurice Stans, Hugh Sloan, Donald Segretti, James McCord, Robert Mardian, Jeb Magruder, Egil Krogh, Herbert Kalmbach, H. R. Haldeman, L. Patrick Gray, John D. Ehrlichman, Anthony Ulasewicz, etc., etc., etc., sat under the glare of camera lights and told their stories. G. Gordon Liddy refused to furnish information. From former Cabinet Secretaries, to former high White House aides to President Nixon, to former officials to CREEP, to former policemen, the threads of the tangled web finally wove themselves into a television pattern for the home screens.

In its *Final Report,* of June 1974, the committee described the self-imposed rules it had followed:

> The character of the committee's hearings resulted from considerable planning and a basic philosophy. The committee, aware of the gravity of the national scandal it was investigating and the fact that its activities would be highly publicized, was determined to present dignified, objective hearings. It recognized that the ultimate impact of its work depended upon obtaining and keeping public confidence.

> In part for these reasons, the committee resisted calling so-called "big name" witnesses at the beginnings of its hearings. The committee and staff wished to present a careful presentation of the evidence establishing a foundation for the later testimony that implicated high government and campaign officials. Early witnesses of lesser stature that enabled the public to understand the context in which the Watergate affair unfolded were essential. . . .
>
> The committee's interest in televised hearings was not to obtain publicity for publicity's sake. The facts which the committee produced dealt with the very integrity of the electoral process; they were facts, the committee believed, the public had a right to know. . . . The ability to read about the hearings in the printed media was not sufficient. The full impact of the hearings could only be achieved by observing the witnesses and hearing their testimony.[3]

Obstructionist tactics devised by, or in behalf of, President Nixon did not stop the processes of mass education. Many of the pro-Nixon maneuverings only inflamed a public sense of betrayal by the Chief Executive. Often, the media—print, radio, and television—made the blundering of the White House more emphatic by playing up all ramifications. For example, Special Prosecutor Archibald Cox made public, in a televised press conference (October 20, 1973), his decision to maintain his demand for the famous White House tapes despite Nixon's refusal to surrender them to him or to the courts or to the congressional investigating committees. A few hours later, Cox was fired by the President and his office was abolished. In the play-out of what is now known as the "Saturday Night Massacre," Attorney General Elliot Richardson and Deputy Attorney General William D. Ruckelshaus also left the administration, resigning in protest after refusing to accept the White House order that Cox be fired on their authority. All segments of the press worked on that story. Within hours Cox, Richardson, and Ruckelshaus were seen by the public as heroes of the law. In a matter of days they were also regarded as pillars of civic morality. It was a time when defenders of righteousness were badly needed, and the public warmed to them when they appeared. "It is," editorialized the *New York Times* on April 22, 1973, "a time of shoddy practices and compromised standards, of rampant cynicism and a get-rich-quick philosophy, of sleazy political methods and an anything goes popular culture.'' [4]

There was very little of the sleazy connected with the work of the House Judiciary Committee's investigation into the propriety of voting impeachment articles against Richard Nixon. Chairman Peter W. Rodino (Democrat, New Jersey) and the thirty-five committee members carried out one of the most difficult constitutional assignments in an orderly and dignified manner. The televised committee hearings restored faith in government across the land. The President's defenders had their full say. On the whole, there was little grandstanding. Those who favored bringing the President to Senate trial, those who hesitated at the contemplation of the drastic step, and those who steadfastly denied that the evidence was conclusive all held the view that their duty to sit in judgment was odious but unavoidable.

There were contrasts, to be sure, made quite clear to television viewers especially. Chairman Rodino's quiet determination to carry out the House mandate fairly was reassuring. Charles Sandman, Jr. (Republican, New Jersey), was flamboyant in his approach and unveering in his refusal to believe that charges had been proved. The earthy stories which managed to get William Hungate's (Democrat, Missouri) points over forcefully were different in tone from Tom Railsback's (Republican, Illinois) serious contemplations as he struggled with his conscience and made known his views about duty. Father Robert F. Drinan (Democrat, Massachusetts) was dour but formidable on the infrequent occasions when he outlined the facts and their significance. Jerome Waldie (Democrat, California) could hardly keep check on his anger toward the President for betraying public trust. Barbara Jordan (Democrat, Texas) spoke with such eloquence that her message sounded like a call to freedom. Joseph J. Maraziti (Republican, New Jersey) made little dramatic impression. His vote, like those of the others, was what ultimately counted.

Three articles of impeachment were passed between July 27 and July 29, 1974. The first charged that President Nixon had ``prevented, obstructed and impeded the administration of justice.'' The second charged that he had ''repeatedly engaged in conduct violating the constitutional rights of citizens, impairing the due and proper administration of justice and the conduct of lawful inquiries, or contravening the laws governing agencies of the executive branch and the purposes of these agencies.'' The third charged that he had ''failed without lawful cause or excuse to produce papers and things as directed by duly authorized subpoenas issued by the Committee on the Judiciary of the House of Representatives . . . and willfully disobeyed such subpoenas.'' [5]

The media brought Nixon down as much as the articles of impeachment. Relentlessly they kept with the story of Watergate, making all the administration's ploys clearly evident. Perhaps their doggedness in pursuing the story is partially attributable to some embarrassment at their past tendency toward timidity in criticizing the domineering Nixon. For illustration, in the election campaign of 1972 television networks drew back from certain types of political coverage. There was an ''almost complete disappearance of television political specials'' between Labor Day and Election Day. CBS, which had presented an average of seven such specials per presidential campaign since 1960, offered only two in 1972. Going against the editorial tide that year, Walter Cronkite, on his CBS evening news program, devoted much time to the Watergate story, the controversial wheat deal engineered by the administration with the Soviet Union, and other stories of that type. On October 27, 1972, he devoted the bulk of the time available on the program to the Watergate story, and in the last week of the campaign he presented six reports on Watergate, for a total of forty-one minutes of coverage of ''The Candidates and the Issues.'' [6] Cronkite, the most popular anchorman in the history of television, could use enormous prestige to be merely a personable news editor and news announcer. Happily, when vital issues are at stake, he proves himself to be a direct intellectual disciple of the late Edward R. Murrow. Murrow took on many causes; Cronkite takes on very few. He does know, however, when courage and professionalism are synonymous.

COSTLY MEDIA POLITICS AND CAMPAIGN REFORMS: NEW CAMPAIGN RULES OF 1971

The Watergate scandals forced new congressional interest in reform of campaign and election laws. A majority of the public had been growing increasingly angered by the daily relevations from investigatory committees that "slush funds" had been made available to finance unwholesome and illegal activities; that corporations and private individuals had contributed enormous amounts of money, apparently to cushion their own nests later with the help of key politicians; that no accounting of much political funding had been made in accordance with corporate tax, income, or election financing laws.

If money is the root of political evil, then controls over its flow are necessary. Reliable estimates show that the amounts spent to win elective office in the United States have risen staggeringly. Since 1952, when candidates expended about 140 million dollars, there has been steadily upward climb, with the figures totaling (approximately) 155 million dollars in 1956, 175 million in 1960, 200 million in 1964, and 300 million in 1968, and with more than 400 million dollars projected before the 1972 campaign began.[7] Exactly how much was spent that year, legally *and* illegally, can only be conjectured.

Members of Congress thought they had done much to control the flow of political dollars just prior to the 1972 election. The legislature had replaced the ineffective Corrupt Practices Act of 1925 with a new law providing the first basic campaign reforms in the nation's history. In the opinion of Senator John O. Pastore (Democrat, Rhode Island), the Federal Election Campaign Act of 1971 was "in terms of its impact on our democratic system and its potential for good . . . a major landmark in the entire history of our legislative process." [8]

That law controlled the amounts candidates for Congress and the Presidency could spend for advertising via television, radio, newspapers, magazines, billboards, and automatic telephones. Each candidate was limited to 10 cents per voter. A maximum of 60 percent of the media expenditures could be used for broadcasting. During the last forty-five days of a primary election and the last sixty days of a general election, broadcasters had to sell advertising time and space to the candidates at the lowest unit rates prevailing. Broadcast stations were required to sell "reasonable amounts of time to candidates."

Candidates and their political committees had to report all contributors who gave or loaned more than $100 and show who received the monies. The candidates themselves and their families were limited in contributions to their own campaigns ($50,000 for presidential or vice-presidential efforts, $35,000 for Senate races, $25,000 for campaigns for the House of Representatives). Continued in force was the "equal time" provision of the Communications Act of 1934. The secretary of the Senate, the clerk of the House, and the United States comptroller general were to receive all the reports from candidates.

Senator Pastore had hoped that Congress would see fit to establish an independent "Federal Elections Commission" to receive the reports and police compliance, but that was not voted in by the joint conference committee which could cut provisions. And there were other shortcomings, including important loopholes such as exclusion of the "costs of producing materials [and] payments to pollsters, media experts, film makers, and researchers." In addition, "Contributions of computer time, postage meters, manpower, etc., . . . were not required to be listed on disclosure reports." [9]

AFTER THE DELUGE: THE FEDERAL ELECTION CAMPAIGN ACT OF 1974

Obviously, the 1971 reforms failed to prevent the political deluge of corrupt activities which the Watergate affair represents to the American people. Nixon, who explained away his supplementary expense fund in September 1952, could not satisfy his critics in 1972. Despite his repeated use of the media to assure the public that all was right and proper, there was to be no repeat of the effects produced by his dramatic television appearance on the evening of September 23, 1952, when he said, ". . . and I want to make this particularly clear, that no contributor to this fund, no contributor to any of my campaigns, has ever received any consideration that he would not have received as an ordinary constituent." In question in 1952 were the uses made of a fund totaling $18,235 and the retention of Nixon as the vice-presidential candidate of the Republican party.[10] In question during the spring of 1974 was a pattern of corrupt behavior which was well financed and woven by men drunk with power.

It is ironical that only two hours before President Nixon resigned on August 9, 1974, the House of Representatives voted, 355 to 48, to pass a new campaign control law, amending the historic legislation of 1971. The next day, compromise was reached with the Senate on the final version to be sent to the President. President Ford signed the Federal Election Campaign Act of 1974 on October 15 of that year.

That law sets new limits on campaign donations from private individuals and groups. It also controls expenditures. It improves reporting requirements. It establishes the independent Federal Elections Commission for supervision and enforcement. And it provides for public financing of presidential elections. This complex legislation, so directly an outgrowth of the scandals of 1972, will be tested in 1976, the bicentennial anniversary of American independence.

Highlights of the congressional reform of 1974 follow.

Funding and Expenditures

Public funding of a Presidential Election Campaign Fund encouraged but did not satisfy many reform groups that had urged public financing for congressional elections as well. The original Senate bill had included provision for such financing, but in the face of strong House opposition the House-Senate conference committee deleted that aspect of reform in the final compromise legislation. The Center for Public Financing of Elections—one of the more prominent of the some thirty private labor, business, church, citizen-action, and civil rights groups spearheading the drive for changes in our election procedures—expressed the view that "the next Congress must continue the fight to end the unholy alliance between money and politics. Public financing of all political campaigns is an idea whose time will come."

The formulas for public financing of presidential races apply to monies made available since Congress passed the Revenue Act of 1971, which provided a checkoff option on federal income tax forms (Internal Revenue Service Form 1040, line 8). Taxpayers have been able to designate a $1 contribution ($2 on a joint return) for this purpose. Substantial amounts have been built up from very small contributions. As of October 1974 the checkoff fund contained 30.1 million dollars, and it is estimated that by 1976 there will be at least 64 million dollars in the fund. Most probably there will be considerably more. For 1976 expenditures, 44 million dollars will be set aside to finance the general election and the party conventions. There will also be funding for the primaries. According to the new rules, a maximum of 45 percent can go to candidates of one political party, and no candidate is eligible for more than one-fourth of available public funds for primaries.

Candidates are not obliged to use public money. All are allowed to raise funds privately. However, if any candidate relies on private contributions, individual donations are limited to $1000 and organizational contributions to $5000. No individual is allowed to contribute more than $25,000 for all federal campaigns for an entire campaign period. The candidates themselves are restricted in the amounts they may donate to their own campaigns: Presidency, $50,000 for the entire campaign; Senate, $35,000 for the entire campaign; House of Representatives, $25,000 for the entire campaign.

Presidential candidates who are *privately funded* are restricted to expenditures of 10 million dollars for the primary campaign and 20 million dollars for the general election. They are allowed an additional 20 percent for fund raising in each campaign. Also, candidates in a presidential primary may not spend more than twice what a candidate for the Senate is permitted.

Presidential candidates opting for *public funding* have to follow the rules for the primary, the convention, and the general election. To be eligible for primary campaign support, via federal matching of private contributions, each candidate must qualify by raising a minimum of $100,000 ($5000 in each of twenty states, at least). The U.S. Treasury then provides for matching individual private contributions up to $250. Public money is available for party conventions, to a limit of 2 million dollars. Major parties qualify automatically. Minor parties can receive amounts prorated to their proportion of votes in the past or current election. For general election uses, 20 million dollars is available. A major party nominee is automatically qualified for such funding in the general election. Again, the minor party candidates are eligible on the basis of a prorated formula related to past or current votes received. In any case of full public funding, private contributions are not allowed.

Those running for the Senate or for representative-at-large seats may spend $100,000 on primary elections *or* 8 cents multiplied by the voting-age population, whichever is greater. In general elections they are restricted to $150,000 or 12 cents multiplied by the voter population, whichever is greater.

Would-be members of the House of Representatives are allowed $70,000 in each of the elections—primary, general, and runoff.

Candidates are allowed to spend 20 percent more than the set limits to meet the costs of fund raising.

The 1971 law's limits on media spending were repealed.

Reporting and Disclosure

Each candidate must set up one central campaign committee which must report all contributions and expenditures. These central committees are required to have specific bank depositories and must file full reports with the Federal Elections Commission ten days before and thirty days after every election. Also, reports are required within ten days of each calendar quarter unless income or outgo is less than $1000.

Private organizations are also bound by new disclosure regulations. Congressman Wayne Hays outlined the essentials to his colleagues before they voted on the conference version of the bill:

That section requires organizations that buy space or time in the mass media or utilize such means as mailings to nonmembers in order to influence the outcome of an election or to state a candidate's position on any public issue, voting record or other official act, to report in essentially the same manner as a political committee.

Enforcement

Under this law, a six-member Federal Elections Commission is responsible for administration of the election law, for the public financing program, and for primary civil enforcement. The selection plan devised is rather unusual. The President, the Speaker of the House, and the President Pro Tempore of the Senate each nominate two members (of different political parties), who are subject to confirmation by both houses of Congress. At the time of appointment no nominee may be employed by, or an official of, any branch of government. (On January 30, 1976, the Supreme Court ordered restructuring of the Federal Election Commission.*)

---

* As this study is going to press, late changes in the law affecting campaigns for Congress and the Presidency are being made. It is unfortunate that the fundamental reforms so long advocated will have to take effect in the campaign of 1980, if then. At this writing, the confusion and unfairness created by the candidates' need for tremendous campaign funds in this media age are only compounded.

On January 30, 1976, the Supreme Court ruled that the Federal Election Commission had to be restructured or cease most of its activities within thirty days. Because the majority of Commission members had been named by officials of Congress rather than by the President, the Court ordered congressional revisions in keeping with constitutional precedents. Congress took its time despite the needs of candidates during the very active presidential primary season. President Ford signed the new bill on May 11, thus reviving the Commission, and notified Congress to expect his nominations shortly. He assented with reluctance. In particular, he was opposed to the new provision that either house of Congress could veto a regulation passed by the Commission. Ford asked Attorney General Levi to look into the constitutionality of that part of the legislation.

Other changes mandated by Supreme Court decision altered the situation substantially. For example, candidates for congressional seats will not be subjected to any spending limits. Presidential candidates who refuse federal funds can also spend as much of their own or their immediate families' money as they desire before the party convention.

Referring to *independent* efforts of private citizens, the Court allowed the expenditure of any amount to defeat or help to elect a candidate. For individuals contributing to a candidate in a primary or a general election, the $1000 limits were upheld.

On the relevance of the First Amendment guarantees of freedom of expression, the Court concluded:

The secretary of the Senate and the clerk of the House serve as ex officio, nonvoting members of the Commission and are to act as custodians of reports for congressional candidates.

The commissioners are to serve full-time six-year staggered terms of office. The chairmanship of the commission will rotate yearly.

Responsibilities of the commission include the following: receiving campaign reports, making rules and regulations (which Congress can review within thirty days), maintaining a cumulative index of reports filed and not filed, making both regular and special reports to Congress and the President, and serving as an election information clearinghouse. The commission is authorized to give "advisory opinions, conduct audits and investigations, subpoena witnesses and information, initiate civil proceedings for relief." Any criminal violations are to be referred to, and prosecuted by, the Justice Department.[11]

CONCERNS ABOUT THE 1974 REFORM

There is good reason to be hesitant about shouting out hurrahs, acting on the premise that the bad times are behind us. For one thing, this reform applies only to the candidates themselves and has no

---

> The ceiling on personal expenditures by candidates on their own behalf . . . imposes a substantial restraint on the ability of persons to engage in protected First Amendment expression. The candidate, no less than any other person, has a First Amendment right to engage in the discussion of public issues and vigorously and tirelessly to advocate his own election and the election of other candidates.

The Court referred to old and sagacious conclusions of Justice Louis Brandeis: "Publicity is justly recommended as a remedy for social and industrial diseases. Sunlight is said to be the best of disinfectants; electric light the most efficient policeman." (See "Excerpts from Supreme Court's Decision on Federal Election Campaign Act," *New York Times,* Jan. 31, 1976.)

The legislation reactivating the Federal Election Commission permits private corporations much leeway in raising money for political contributions from their executives and rank-and-file employees. Labor unions are now required to report to their members on expenditures for communications or contributions for favored political candidates. In 1976 only the tip of this iceberg, made up of corporate and union funds, will appear; in 1978 and 1980 such funds could alter the course of the ship of state.

impact upon the pressure groups which have been and probably still are eager to use money to attain their selfish ends, not necessarily connected to the public good. If the campaign controls are to have a fighting chance, Congress must get down to the business of drafting a substitute for the Federal Regulation of Lobbying Act of 1946, which has become virtually ineffective as a control over the abuses perpetuated by unbridled pressure groups in this country. Interpretation of that law by the Supreme Court in 1954 so modified its applicability that astute observers do not take it seriously. In that year the Court ruled that lobbying was to be defined as direct communication with Congress. Obviously, then, lobbying activities aimed at the White House, the federal bureaucracy, and Cabinet secretaries have been exempt from regulation.

There is also a serious legal challenge to the Federal Election Campaign Act of 1974, filed in the District Court of the District of Columbia by Senator James L. Buckley of New York, former Senator Eugene McCarthy of Minnesota, Congressman William Steiger of Wisconsin, and other "concerned citizens." * They contended, in January 1975, that "substantial constitutional questions" had to be raised.

This challenge is interesting for several reasons. First, allied together are liberals and conservatives of the political spectrum. Second, most of the key provisions of the Federal Election Campaign Act of 1971 as amended in 1974 are attacked on constitutional grounds. One broad charge in the complaint (No. 63) is illustrative of others:

> The FECA Amendments deprive one or more of the plaintiffs of their rights in that the method of appointment of the Federal Election Commission violates the constitutional separation of powers and discrimi-

---

* Parties before the U.S. District Court, District of Columbia, in civil action U.S. 750001, in the "Complaint for Declaratory and Injunctive Relief," include Senator James L. Buckley of New York; Eugene J. McCarthy, formerly senator from Minnesota; Congressman William A. Steiger of Wisconsin; Stewart R. Mott; the Committee for a Constitutional Presidency; the Conservative party of the State of New York; the New York Civil Liberties Union, Inc.; the American Conservative Union; and Human Events, Inc.

nates invidiously against them in violation of the Due Process Clause of the Fifth Amendment to the Constitution.

Plaintiff Buckley notes:

The expenditure of funds is essential to his success in conveying his ideas and position on public issues to voters and potential voters. To that end plaintiff desires to solicit and receive contributions larger than $1,000, employ personal funds or the funds of his immediate family as available and necessary, and to spend whatever amounts are available and necessary to convey his views to voters and potential voters in a meaningful way.

At the heart of the challenge is this complaint (No. 60):

The FECA and the FECA Amendments violate the constitutional rights of one or more of the plaintiffs in that they limit the expenditures of national committees of a political party in the case of campaigns for federal office in violation of the rights of freedom of speech and association guaranteed by the First Amendment to the Constitution and the Due Process Clause of the Fifth Amendment to the Constitution.[12]

So, control over spending is tied by plaintiffs Buckley, McCarthy, et al., to freedom of expression. They fear that the new law encroaches upon essential private rights and suspect that this is the opening wedge of a drive to make those seeking political office subservient to the government in power. It would be satisfying to conclude that their claims are not worthy of the deepest study because their approach is obstructing *reform,* but that would not be a valid judgment. The reforms of 1974 were passed by a Congress concerned about proving that it could provide a righteous alternative to Watergate-type scandals. In short, the legislation is essentially reactive.

Limits on total amounts spent do not ensure fair uses of media in terms of purposes or content of messages. Repeal of the 1971 law's complicated formulas regulating proportions of total funds permitted for media expenditures, without an appropriate substitute, may increase political pugnacity in elections.

SIGNS OF THE TIMES: MEDIA POLITICS AND ELECTIONS

Politicians win or lose elections today primarily on the basis of their successes or failures in the area of media management. Sometimes this management is described as ''scientific politics'' or ''management politics.''

Note that it is not suggested that politicians necessarily control or even fully understand this new *science* or presentation and display. Indeed, one of the grave problems is that a new category of experts has imposed itself upon the political scene, claiming that traditional political solicitations which centered on the person seeking office are outmoded and unprofessional.

John S. Saloma and Frederick H. Sontag, in their careful and basic study *Parties: The Real Opportunity for Effective Citizen Politics,* observe that ''the professionalization of the parties will have the added effect of concentrating new political resources in national and state party committees with wide discretionary authority in hiring trained professionals who often have little or no public visibility.''[13]

This new trend has at least one certain result. There is a turning away from political amateurs whose enthusiasm or talent does not match the images which the media managers decide will *sell*. If the one proven group of experts in a political campaign consists of media managers, fighting with each other *through* the candidates they maneuver, then the politicians are virtually superfluous until they learn how the electorate voted.

Obviously, media expertise is not necessarily political expertise. We are near that dangerous flash point in this country when we will have to decide whether we want the best candidates or the best media stars. Media managers, who label themselves political consultants, take pride in how instrumental they are in determining outcomes on Election Day.

One of the best of the new breed is Joseph Napolitan, founder of the American Association of Political Consultants, who has handled leading politicians such as Milton Shapp, Hubert Humphrey, and Mike Gravel. His book *The Election Game* conveys the vivid impression that he sees political campaigns as media jousts managed around, and perhaps despite the influences of, old-fashioned politics. Here is the essential Napolitan: "First, define the message the candidate is to communicate to the voters. Second, select the vehicles of communication. Third, implement the communication process." [14] And, "The new politics is the art of communicating a candidate's message directly to the voter without filtering it through the party organization." [15] He suggests that elections are more important than conventions; and that politicians ought not to forget that it is the most telling slogans, film clips, and radio spots, and the mass letter-mailing campaigns, that are important. To such political consultants the vital factors are these: how to buy media time or space scientifically; how to *pace* your candidate so that he delivers the telling arguments at the crucial point in the race; how to order film makers' products and tailor them to television, in keeping with the entertainment and news presentation trends of the day; and how to catalog the client's assets and liabilities, and then orchestrate his ever-greater public appeal by accentuating the positive and minimizing the negative.[16]

Realistically, no blanket attack on political consultants can or should be mounted. What deserves scrutiny is the separation of media and politics with the emergence of two distinct types of entrepreneurs working in tandem in campaigns. Are we getting closer to the day when the consultants will no longer need the politicians? Al Smith once said that he could beat one of his opponents "on a Chinese laundry ticket!" Suppose that cocky attitude were adopted by political consultants: at least one of their number might boast that he could beat *his* opponent with a political robot.

Political consultants are, to be sure, very useful in an era of complex communication, but it is high time that some regulation of their activities were considered. At the very least, an enforceable code of conduct should be demanded. At present, one can have no real confidence that there is even an agreed-upon code of ethics covering political campaign management.

On this matter, as on the other vital subjects before all advocates of freedom, success is not so important as consequences.

If political parties are to continue to function as the primary sources of nominees for public office and as major organizational boosters of political policies, they will have to develop in keeping with the vibrant forces now building and straining the collective society. Even now, in most pronouncements about the genius of basic American political structures, one notices more faith in the processes of this democracy than actual evidence of effective working mechanisms. The aged, the young, the disenfranchised, the demanding minorities too long excluded from real participation, the *independents,* the members of minor parties, the ideological radicals, the theoretical avant-garde—all push and pull, raising the possibility that at an uncertain point the structure of government will fail to hold. That is why reform of media politics is so essential.

Perhaps the political consultants, who now devote so much of their time to elections, might turn some of their attention to those who are losing in contests for which no paper ballots or electronic voting machines are made available. Perhaps they could design media manipulations for the long periods of time *between* elections, to help those in nursing homes. The public press has been informative about swindles and scandals perpetrated upon the aged and infirm by operators who milk governmental programs such as Medicaid and Medicare, as, for example, with observations such as this:

> The opportunity, once perceived, has drawn into the nursing home field con men and manipulators whose skills and imagination put them into a class apart from those operators whose stealing is limited to kickbacks from pharmacies and doctors. Few of these newcomers have any experience in nursing homes or in any other aspect of the health field. All they know is how to make money, and they sense that the nursing home is a good place to use their talent.[17]

Maybe there should arise a new breed of political consultants—*investigative political consultants*—as a natural outgrowth of investigative reporting and a refreshing turn away from traditional bonanzas created for those pressure groups that can pay well.

TELEVISION AND POLITICS: REASONS FOR PESSIMISM

To an appreciable degree, the mass media have changed the patterns of political party loyalties in this country. One factor accounting for this development is the feasibility of instant coverage of events by the electronic media. On the whole, such events are rarely interpreted thoughtfully. Walter Cronkite has sat squarely in front of a great many large-screen projections of Vietnam, rattling off the *news* about this battle or that strategy meeting, unable to relay information of real significance. Daniel Rather, Douglas Kiker, Catherine Mackin, Roger Mudd, Daniel Schorr, Irving R. Levine, Rebecca Bell, Herbert Kaplow, and a host of other famed television reporters have stood in front of the White House or the State Department or whatever, using buildings as backdrops and mounthing monumental nothings about events in the news. Sometimes the obvious television story appears to be other than what is described, in sonorous verbiage glossing over superficial analysis, by the reporters in front of the White House. Sometimes the television audiences look behind the reporter's shoulders, up to the front portico of the President's residence, and contentedly note that the great lantern over the front door casts delightful rays over the lawn or that the splendid building does not need painting. "The TV reporters," says Timothy Crouse in *The Boys on the Bus,* his fine recapitulation of how the campaign press corps works (and does not work), "were the direct descendants of Nathaniel Currier and James Ives, the pioneers of American pictorial journalism." He observes, correctly, that "above all else, TV Reporters were trained to search for a good picture." Roger Mudd has said, "The thing with television is that everybody's a high priced communicator and nobody can really communicate."

That being essentially obvious to anyone who watches most television reporting, it follows that, instead of helping us all to understand political issues, the electronic media help us to recognize political personalities. The politicians are created in the reportorial image today. The reporters are created in the entertainment molds.

too often. It also follows that what is considered news deals less and less with what men and women are saying. Coverage increasingly relates to their present success or potential prospects as contestants in elections. Governor George Wallace of Alabama is a good case in point. For many years, reporters, for print and tube, have repeated the question: Will he run for President? Droned into our ears through the 1960s and up to 1976, at least, that question drowns out information about what he means to do, stands for or against, or believes. The same treatment is given Senator Edward Kennedy and Ronald Reagan, former governor of California. Even the politicians are unable to deflect the press corps from its chosen preoccupation. The reporters are now so self-confined that they consider it a terrible professional fate to be assigned to a candidate who is not a front-runner. "If he can hang on to a winner through the primaries, he will probably be assigned to follow him through the fall election—perhaps all the way to the White House." [18]

It is becoming clearer to astute political reporters like David Broder of the *Washington Post* that "television has seemed to make one of the party's old functions irrelevant—that of serving as a bridge between the candidate or officeholder and the public." Broder "is not optimistic about the prospects of reviving responsible party government in the near future. The momentum of current trends, the drift of the public mood seem to me to point in the opposite direction: toward the further fracturing of the already enfeebled party structure in this decade." He believes that "not only are voters splitting their tickets and moving back and forth from election to election, but their perception of party differences is growing visibly weaker." Broder entitles his book *The Party's Over.* [19]

The media circumstances of our times help us to understand, for example, why there are such election landslides as that which renewed Richard Nixon's Presidency in 1972. People who used to be called independent voters are now ticket-splitters. (A ticket-splitter is "an involved voter who rationally splits his ballot from President to state legislative races in a single election and tends to split his vote in successive elections.") Large numbers of previous party loyalists are also now converted to ticket-splitters.

There are direct relevancies in the research findings that the ticket-splitter gets his or her primary information largely from television. Walter De Vries and V. Lance Tarrance, in their brilliant study *The Ticket-Splitter,* report on a 1970 gubernatorial election survey (with 809 respondents) conducted in Michigan. When asked to rate thirty-five variables that were or might have been important to them as they made their political decisions, the print media came out poorly (with newspaper editorials and stories the only categories to go over five points on an eleven-point scale). Undecided voters and ticket-splitters alike gave high ratings to the audiovisual media. The appreciation of both groups is attested to by the result that differences did not exceed 0.3 percent: television newscasts were ranked at 6.7 percent by undecided voters and 6.8 percent by ticket-splitters; television documentaries and specials, at 6.5 percent by undecided voters and 6.6 percent by ticket-splitters; television editorials, at 5.7 percent by undecided voters and 5.6 percent by ticket-splitters.

De Vries and Tarrance feel that William Milliken, one of only two midwestern Republican governors who won in 1970, succeeded because he overcame the odds against him by gearing his political campaign to reach the ticket-splitters of Michigan. Television was extensively used in his behalf, and every device associated with regular news presentations was incorporated into his spots. Obvious propaganda was soft-pedaled, and "newsy" information was stressed. To illustrate, nine television spots each opened with a black-and-white photograph "emphasizing the graphic starkness" of each problem presented. Governor Milliken was then shown "demonstrating his understanding . . . and addressing himself to its solution." [20]

Election issues, it is clear, are coming to resemble television news issues: result—appearance is more important than argument. Politicians, unfortunately, have demonstrated an increasing tendency to replace concern for policies with a predilection for posturing. Indeed, there is so much posturing that our traditional democratic beliefs are being dangerously diluted and exchanged for short-term political preferences induced by television's political commercials.

For convenience, some researchers have divided voters into three groups: low-interest, moderate-interest, and high-interest. One of the characteristics of the high-interest segment is the tendency to reply on multiple channels of political information; of the low-interest segment, to rely on one of the mass media, *television*.

According to an especially significant finding in recent research on the 1972 presidential election, levels of information attained by low-interest voters may equal those of high-interest voters. Thus, the kinds and quality of information transferred by the media to voters are more important than the amounts. Television's political advertisements apparently have a strong influence on the beliefs of low-interest voters, possibly to the extent of effecting changes in those beliefs. Moreover:

> On the belief items included in the messages, televised commercials function primarily to reduce voter-to-voter differences in beliefs about the candidates. Among the voters most affected are those who normally would not be interested in the information conveyed, particularly low-interest voters and opposition supporters who tend either to ignore political communications generally or to ignore specific messages. By by breaking down the tendency toward selective exposure, political ads provide these voters more, or different, political beliefs than they would otherwise possess.[21]

Apart from the belief-changing influence implicit in much current political advertising, there are other complications. One is not surprising, in that political consultants draw heavily on the creative traditions of commercial product advertising. Often the winner of an electoral contest succeeds because of good *packaging*. One very successful consultant, Tony Schwartz, who handled the media work of Senator Abraham Ribicoff (Democrat, Connecticut) in the 1974 election campaign, has been described by Joseph Napolitan as "the best in the business" when it comes to political sounds and pictures. Another renowned consultant, David Garth, who managed the media work in Hugh L. Carey's successful race for governor of New York in 1974, considers him "the master" of the tape recorder. Schwartz has concluded that he studies "how to surround the voter with auditory and visual stimuli that evoke feelings that move him to pull our candidate's lever in the voting booth." [22]

Dirty politics, aided by mass media, is the worst problem confronting us. Appeals to rouse racial and religious bigotry are not new in politics. In 1966, 18 percent of the complaints addressed to the Fair Campaign Practices Committee (a private organization) dealt with racial and religious issues. Happily, by 1970 the same group received only 3 percent of its complaints on those subjects. But there are many ways to be dirty, unfortunately. In 1970, when Senator Vance Hartke and Richard Roudebush were competing for office, Hartke complained to the FEPC about a questionable television ad. In it, a Viet Cong soldier was handed a rifle by someone while a voice-over said that Senator Hartke's vote favoring trade with Communist countries resulted in the provision of guns used to kill and wound American boys. The voice-over concluded with, "Isn't that like putting a loaded gun in the hands of our enemies?" [23]

David Broder summed up all the essentials when he recalled a luncheon conversation he had, after the 1964 election, with people who had helped devise the successful media strategy for President Johnson. They were joyously describing the ways they had, as Broder puts it, "foisted on the American public a picture of Barry Goldwater as the nuclear-mad bomber who was going to saw off the eastern seaboard of the United States and end everyone's Social Security benefits." One of Broder's luncheon companions tempered his enthusiasm with an ethical comment: "The only thing that worries me, Dave, is that an outfit as good as ours might go to work for the *wrong* candidate." [24]

THE ROADS AHEAD

Several years ago the author traced the influences of television on American political life in the book *Political Television.* [25] After a review of the national campaigns of 1960 and 1964, with emphasis on the presidential preoccupations of the electorate, one conclusion was that good government was somewhat handicapped because of presidential primary and general election "fever," which lasted for at least two years out of every four. In addition, financial burdens on candidates were becoming prohibitive; media imagery was becoming more important than the taking of political positions; and campaigns were becoming ends in and of themselves, as if the objective were something other than government service.

Since that appraisal there has been considerable movement but not much change relative to some of the major problems. Media imagery is of greater importance now, for better or worse. National campaigns are just as long, and fewer ordinary citizens can afford to run for public office because of the crippling expenditures attached to campaigning.

There are some hopeful signs, however. Reforms are being implemented to improve the campaign finance situation considerably. A change of national mood, after the Watergate affair and the Vietnam conflict, indicates that a more reasonable political environment is likely to emerge. The general public is assured that it is entitled to substantial government information and to better protection by a press corps that has learned to be more sensitive to infringements on the Bill of Rights.

One problem persists. On many subjects, we are not realistic as a people. In respect to the most popular medium of communication, television, we do not insist that it be used effectively as an educational instrument. Therefore, many issues of government which have backgrounds in the social, economic, political, and environmental facts of life remain mysteries.

A caustic critic of Franklin D. Roosevelt's New Deal once accused his administration of "playing craps with destiny." Despite all the reforms discussed in these pages, we still seem to enjoy that perilous game. We do not appear to know what the stakes are if politics becomes a segment of show business. Are we not like that grandmother who proudly displayed her infant grandchild to a friend? In response to a very generous remark about the baby, she said, "This is nothing. You should see his pictures!"

It is high time that, as a rule, the mass media concentrated on reality to the extent necessary.

If this democracy is not only to survive but to thrive and be an inspiration to the world, we must never forget that the mass media have unparalleled civic responsibilities. Our politics at home and abroad increasingly depends upon how well they help us to admire and protect real babies, and to see the roads ahead that all will have to travel.

NOTES

CHAPTER 1

1. See *In the Public Interest* (New York: National News Council, 1975), pp. 7–12, 78–81, 113–116.
2. See Lesley Oelsner, "Blackmun Backs a Curb on Press in Nebraska Case," *New York Times,* Nov. 22, 1975; Martin Arnold, "Blackmun Press Club," *New York Times,* Nov. 27, 1975; Tom Wicker, "Blackmun v. Press: A Bad Case," *New York Times,* Nov. 25, 1975.
3. Harold R. Medina, "Omnibus 'Gag' Rulings," *New York Times,* Nov. 30, 1975.
4. See transcript of "Public Enemy Number One," of the "Behind the Lines" Educational Broadcasting Corporation television series. New York air date, Jan. 2, 1975. Mimeographed version of edited script provided by WNET 13, Box 345, New York, 8 pp.
5. Henry Fairlie, "The Lives of Politicians," *Encounter,* vol. 29, no. 2, pp. 23–25, August 1967.
6. Carl J. Friedrich, *The Pathology of Politics* (New York: Harper & Row, 1972), p. 15.
7. See Robert E. Summers and Harrison B. Summers, *Broadcasting and the Public* (Belmont, Calif.: Wadsworth Publishing Company, 1966), pp. 190–191.
8. Alexander Kendrick, *Prime Time: The Life of Edward R. Murrow* (Boston: Little, Brown and Company, 1969), pp. 329, 340.
9. Ibid., pp. 340–341.
10. Evron M. Kirkpatrick, "Toward a More Responsible Two-Party System: Politi-

cal Science, Policy Science, or Pseudo-Science?" *The American Political Science Review,* vol. 65, no. 4, pp. 970–972 (December 1971).

11. Ibid., p. 986.
12. Ibid., p. 971.
13. Christopher H. Sterling, "Some Basic Limitations in Mass News," in David J. Leroy and Christopher H. Sterling (eds.), *Mass News* (Englewood Cliffs, N.J.: Prentice-Hall, 1973), pp. 71, 77.
14. Clifton Daniel, "Responsibility of the Reporter and Editor," in Louis M. Lyons (ed.), *Reporting the News* (New York: Atheneum Publishers, 1968), p. 121.
15. Donald C. Pirages and Paul R. Ehrlich, *Ark II* (New York: The Viking Press, 1974), p. 189.
16. Ibid., pp. 120–126. Also see Barbara Ward, *The Rich Nations and the Poor Nations* (New York: W. W. Norton & Company, 1962), pp. 141–142, for this trenchant summarization:

> Looking at our society I certainly do not feel that it already presents such an image of the good life that we can afford to say that we have contributed all that we can to the vision of a transfigured humanity. Our uncontrollably sprawling cities, our shapeless suburbia, our trivial pursuits—quiz shows, TV, the golf games—hardly add up to the final end of man. We can do better than this. We also have the means to do better. If we do not feel the need there is only one explanation. We no longer have the vital imagination for the task.

CHAPTER 2

1. Oran R. Young, "Intervention and International Systems," *Journal of International Affairs,* vol. 22, no. 2, p. 178, 1968.
2. See Lawrence Dietz, "The Selling of the $2.2 Billion TV Season," *New York Times,* June 23, 1974, sec. 2, pp. 1, 19.
3. Ben H. Bagdikian, *The Information Machines* (New York: Harper & Row, 1971), pp. 296–297.
4. Robin Day, "Troubled Reflections of a TV Journalist," *Encounter,* vol. 34, no. 5, pp. 77–88, May 1970.
5. T. R. Fyvel, "Children of the Television Age," *Encounter,* vol. 35, no. 4, p. 46, October 1970.
6. Daniel Bell, "Sensibility in the 60's," *Commentary,* vol. 51, no. 6, p. 73, June 1971.
7. David Easton, "A Systems Analysis of Political Life," in Walter F. Buckley (ed.), *Modern Systems Research for the Behavioral Scientist* (Chicago: Aldine Publishing Company, 1968), p. 429.
8. See *Time,* vol. 102, no. 16, Oct. 15, 1973.

9. R. Vincent Farace, "Mass Communication and National Development: Some Insights from Aggregate Analysis," in David K. Berlo (ed.), *Mass Communication and the Development of Nations* (East Lansing, Mich.: International Communication Institute, Michigan State University, 1968), chap. 5.
10. Max F. Millikan, "TV and Emerging Nations," *Television Quarterly,* vol. 7, no. 2, p. 31, Spring 1968.
11. For pertinent commentary see Henry S. Kariel, "Creating Political Reality," *The American Political Science Review,* vol. 66, no. 4, p. 1094, December 1970.
12. U.S., Congress, Senate, Committee on Commerce, Subcommittee on Communications, *Violence on Television: Hearings,* ser. no. 93–76, 93d Cong., 2d Sess., April 3–5, 1974, p. 160.
13. Ibid., pp. 25–26.
14. Ibid., pp. 5, 8.
15. Ibid., p. 31.
16. Ibid., pp. 52, 63. For 1974 material, see Franklynn Peterson, "Too Much TV Distorts Reality," *Boston Globe,* Aug. 27, 1974. Quotations relating to later Gerbner research are taken from materials provided to the author by Dr. Gerbner on Sept. 24, 1974. See George Gerbner and Larry Gross, "Cultural Indicators: The Social Reality of Television Drama—Proposal for the Renewal of a Research Grant," n.d., 25 pp. (see pp. 15–16 especially).
17. See Jacques Ellul, *The Political Illusion* (New York: Alfred A. Knopf, 1967), pp. 55–56, 59–60, 103–104.
18. See *Report of the National Advisory Commission on Civil Disorders* (New York: The New York Times, 1968; Bantam Books ed.). Members of the Commission included Otto Kerner, Chairman; Mayor John V. Lindsay, of New York City, Vice Chairman; U.S. Senator Edward W. Brooke; U.S. Representatives James C. Corman and William M. McCulloch; I. W. Abel, President of the United Steelworkers of America; Charles B. Thornton, chief executive officer of Litton Industries; Roy Wilkins, of the NAACP; Katherine G. Peden, Commissioner of Commerce of Kentucky; Herbert Jenkins, Chief of Police, Atlanta, Ga. All titles refer to individual positions at time of appointment.
19. Ibid., pp. 407–408.
20. Ibid., pp. 362–366.
21. Ibid., pp. 369–371.
22. Ibid., p. 383.
23. Ibid., pp. 384–388. For details on Columbia University Program, see Nathaniel Sheppard, Jr., "Minority-Journalists' Program Closes," *New York Times,* Aug. 17, 1974.
24. See Surgeon General's Scientific Advisory Committee on Television and Social Behavior, *Television and Growing Up: The Impact of Televised Violence, Report to Surgeon General, U.S. Public Health Service* (Washington: U.S. Government Printing Office, January 1972); Special Committee on Radio and Tele-

vision of the Association of the Bar of the City of New York, *Radio, Television and the Administration of Justice* (New York: Columbia Univesity Press, 1965); *Violence and the Media: A Staff Report to the National Commission on the Causes and Prevention of Violence, Hearings* of Oct. 16 and 17 and Dec. 18–20, 1968, vol. 9A (Washington: U.S. Government Printing Office, 1968); U.S., Department of Justice, Community Relations Service, Office of Media Relations, "The News Media and Racial Disorders: A Preliminary Report," *Columbia Journalism Review,* vol. 6, no. 3, pp. 3–5, Fall 1967.

25. Ralph Nader, "Environmental Deterioration as Violence," speech given at Sierra Club *Conference on the End of the Earth: What Professionals Can Do to Save the Environment,* Cambridge, Mass., Apr. 25, 1970.
26. See F. Earle Barcus, "Description of Children's Television Advertising," statement before U.S. Federal Trade Commission, *Hearings on Modern Advertising Practices,* Nov. 10, 1971, 13 pp. mimeographed.
27. Surgeon General's Scientific Advisory Committee or Television and Social Behavior, op. cit., pp. 123–124.
28. Ibid., pp. 111–112. For other illustrative background see "Recipe for Fall Kidvid Lineup: Almost All Animation with a Liberal Sprinkling of Learning," *Television/Radio Age,* pp. 31–34, 80–81, (June 25, 1973; "Children's TV: Much Talk, Few Answers," *Broadcasting,* pp. 39–41, Oct. 9, 1972; Michael Putney, "Zoom . . ." Means New Luster for Adventurous WGBH-TV", *The National Observer,* June 3, 1972; Theodore J. Jacobs, "What's Wrong With Children's Television," *The New York Times,* Sunday, December 27, 1970; "FCC Moves Against Children's TV", *Broadcasting,* vol. 80, no. 4 (Jan. 25, 1971); Address by Dean Burch, Chairman, Federal Communications Commission, before the International Radio and Television Society, Sept. 16, 1970, New York City, New York, in *News* published by the Federal Communications Commission. Also, see newsletter *ACT* published regularly by *Action for Children's Television,* 33 Hancock Avenue, Newton Center, Mass. For example, see *ACT,* vol. 4, no. 1 (Spring/Summer 1974). Also, for recent developments on advertising practices, see "NAB's TV Board, by 8-to-4 vote, ratifies code's restrictions on children's ads," *Broadcasting,* vol. 86, no. 23 (June 8, 1974), p. 24; "Children's TV back to a boil among medium's priorities," *Broadcasting,* vol. 86, no. 14 (April 8, 1974), pp. 20–25.
29. See Robert Berkvist, "Misterogers Is a Caring Man," *New York Times,* Nov. 16, 1969.
30. See "Kidvid '72: Quality Replaces Quantity," *Television/Radio Age,* vol. 20, no. 3, pp. 17–19, 60–62, Sept. 4, 1972. "Take a Giant Step" was introduced by NBC television in the 1971–1972 season. Designed for seven- to twelve-year-olds and featuring innovative preparation of individual programs by teams of thirteen- to fifteen-year-olds, it was intended "to help children to clarify their values and make their own value judgements." Through information and entertainment and with a good measure of participation by the young creators, "children will be encouraged to make free choices after thoughtful consideration of

alternatives on topics such as beauty, 'future shock', food, machines, time, migration and the like.'' For fuller summary see David F. Leiss and Lillian Ambrosino, *An International Comparison of Children's Television Programming* (Washington: National Citizen's Committee for Broadcasting, July 1971), pp. 141–143.

31. See Linda Francke, ''The Games People Play on Sesame Street,'' *New York,* Apr. 5, 1971, pp. 26–29. Also, for background, see Theodore J. Jacobs, ''What's Wrong with Children's Television,'' *New York Times,* Dec. 27, 1970; ''FCC Moves Against Children's TV,'' *Broadcasting,* vol. 80, no. 4, Jan. 25, 1971; Address by Dean Burch, Chairman, Federal Communications Commission, before the International Radio and Television Society, Sept. 16, 1970, New York, in *News,* published by Federal Communications Commission. Also see newsletter *ACT,* published regularly by Action for Children's Televison, 33 Hancock Ave., Newton Center, Mass.; for example, see *ACT,* vol. 4, no. 1, Spring/Summer 1974. Also, for recent developments on advertising practices, see ''NAB's TV Board, by 8-to-4 vote, Ratifies Code's Restrictions on Children's Ads,'' *Broadcasting,* vol. 86, no. 23, p. 6, June 10, 1974; ''Children's TV Back to a Boil among Medium's Priorities,'' vol. 86, no. 14, pp. 20–25, Apr. 8, 1974.
32. See C. P. Snow, ''The State of Siege,'' in his *Public Affairs* (New York: Charles Scribner's Sons, 1971), pp. 202–203, 205, 208–210, 220.
33. Michael S. Teitelbaum, ''Not Everyone Can Hear a Population Explosion,'' *New York Times,* Aug. 18, 1974.
34. See *ABCs of Radio and Television* (New York: Television Information Office, June 1971), pp. 3–4.
35. See Marvin Barrett (ed.), *Survey of Broadcast Journalism, 1969–1970* (New York: Grosset & Dunlap, 1970), pp. 23–25.
36. See John J. O'Connor, ''TV: Solid Documentaries from CBS and PBS,'' *New York Times,* June 3, 1974.
37. See John J. O'Connor, ''TV: 'The Palestinians,' '' *New York Times,* June 14, 1974.
38. See Cyclops, ''Three Women Alone,'' *New York Times,* June 10, 1974.
39. See Steven R. Weisman, '' 'Jane Pittman' Wins Emmy for the Best Program,'' *New York Times,* May 30, 1974.
40. See Jack Gould, ''TV: Impact of U.S. Air Attacks on North Vietnam,'' *New York Times,* Jan. 24, 1968. See also Extension of Remarks of Mr. Ashbrook, ''Propagandizing for the Enemy,'' *Congressional Record,* vol. 114, no. 3, pp. H91–H92, Jan. 17, 1968.
41. See ''Remarks by Richard S. Salant, President, CBS News, to Boston University School of Public Communication, Boston, Mass.,'' Apr. 28, 1971, 14 pp. mimeographed. Also see Hillier Krieghbaum, *Pressures on the Press* (New York: Thomas Y. Crowell Company, 1973), p. 31.
42. See Marvin Barrett, ed., *The Politics of Broadcasting, 1971–1972* (New York: Thomas Y. Crowell Company, 1973), pp. 22–25.

43. See "Remarks of Richard S. Salant, President, CBS News, to the Boston University School of Public Communication," Apr. 28, 1971, p. 4 (Boston, Mass.), 14 pp. mimeographed.
44. Nicholas Johnson, *How To Talk Back to Your Television Set* (New York: Bantam Books, 1970), pp. 154–155.
45. Wilbur Schramm, "Communication and Change," in Daniel L. Lerner and Wilbur Schramm (eds.), *Communication and Change in Developing Countries* (Honolulu: East-West Center Press, 1967), pp. 19, 27–30.
46. Wilbur Schramm, Philip H. Coombs, Friedrich Kahnert, and Jack Lyle, *The New Media: Memo to Educational Planners* (Paris: United Nations Educational, Scientific and Cultural Organization, 1967), p. 96.
47. Harry G. Schwarz, "America Faces Asia: The Problem of Image Projection," *The Journal of Politics,* vol. 26, no. 3, pp. 532–549, Aug. 1964.
48. Richard R. Fagen, "Relation of Communication Growth to National Political Systems in the Less Developed Countries," *Journalism Quarterly,* vol. 41, no. 1, pp. 87–94, Winter 1964.
49. Ithiel de Sola Pool, "Communication and Development," in Myron Wiener (ed.), *Modernization: The Dynamics of Growth* (New York: Basic Books, 1966), pp. 102, 107.
50. See *New Educational Media in Action: Case Studies for Planners* (Paris: UNESCO, International Institute for Educational Planning, 1967), vol. 1, pp. 34–37.
51. See George Comstock and Nathan Maccoby, *The Peace Corps Educational Television Project in Colombia: Two Years of Research* (Stanford, Calif.: Institute for Communications Research, Stanford University, November 1966), vol. I, pp. 2–16, 23, 45–46, 46–53. On the Jamaica experiences, Dr. John S. Clayton, senior ETV adviser, USAID/Jamaica, wrote a final report for his agency in Feburary 1968; for a view of the El Salvador situation, see Robert Schenkkan, Thomas Livingston, Vernon Bronson, and William Kessler, *The Feasibility of Using Television for Educational Development in El Salvador* (Washington: National Association of Educational Broadcasters, July 1967).
52. See President's Task Force on Communications Policy, *Central Staff Working Paper on the Use of Telecommunications by Less Developed Countries,* Washington, Sept. 12, 1968, pp. 94–95. Another inspection of educational priorities assigned television in Western Europe, Canada, Australia, Japan, and the United States is provided by David Fleiss and Lillian Ambrosino in *An International Comparison of Children's Television Programming* (Washington: National Citizens Committee for Broadcasting, July 1971). See p. 75 for a sample conclusion based on comparative analysis of different national systems. We can take little solace from this observation:

> As long as there is not a single weekday afternoon network children's program, and as long as 16 minutes per hour are devoted to advertising on children's programs, American children's television will remain inferior to children's television in Western Europe, Canada, Japan and Australia.

CHAPTER 3

1. *Concurring opinion of Mr. Justice Brandeis in Whitney v. California,* 274 U.S. 357 (1927), pp. 375–376, 71 L. Ed. 1095, pp. 1105–1106.
2. Samuel D. Warren and Louis D. Brandeis, "The Right to Privacy," *Harvard Law Review,* vol. 4, no. 5, p. 196, Dec. 15, 1890.
3. James Reston, "Advice to College Freshman," in his *Sketches in the Sand* (New York: Alfred A. Knopf, 1967), p. 158.
4. "Text of Hutchins Speech before Society of Newspaper Editors," *New York Times,* Apr. 22, 1955. The members of the Hutchins Commission of 1947 constituted a highly distinguished group. Those cognizant of American developments in the fields of law, economics, philosophy, literature, political science, religion, anthropology, education, and history will recognize the eminence of such as Zechariah Chafee, Jr., John M. Clark, Harold D. Lasswell, William E. Hocking, Archibald MacLeish, Charles E. Merriam, John Dickinson, Reinhold Niebuhr, Robert Redfield, Beardsley Ruml, Arthur M. Schlesinger, and George M. Schuster. Foreign advisers included John Grierson, formerly general manager of Canada's War Information Board; Hu Shih, former Chinese Ambassador to the United States; Jacques Maritain, of France; and Kurt Riezler, then professor at the New School for Social Research.
5. Hillier Kriegbaum, *Pressures on the Press* (New York: Thomas Y. Crowell Company, 1972), p. 222. For the broader statement, see Commission on Freedom of the Press, *A Free and Responsible Press* (Chicago: The University of Chicago Press, 1947), pp. 117–118.
6. "Text of Hutchins Speech before Society of Newspaper Editors," op. cit.
7. See "The Hutchins Report: A Twenty Year View," *Columbia Journalism Review,* vol. 6, no. 2, pp. 5–20, Summer 1967.
8. John Tebbel, "What's the News?," *Saturday Review,* vol. 53, no. 11, pp. 111–112, Mar. 14, 1970.
9. See Les Brown, "Livelier, Longer TV News Spurs Hunt for Men, Material," *New York Times,* Apr. 22, 1974.
10. See Cyclops, "No Network Blazers, Star Reporters or Programmed Laughs—Just News," *New York Times,* Sept. 15, 1974.
11. Sir William Haley, "Where TV News Fails," *Columbia Journalism Review,* vol. 9, no. 1, pp. 7–11, Spring 1970.
12. See the film by director-editor Alain Resnais, *Nuit et Brouillard,* produced by Como-Films, Argos-Films, and Cocinor, Paris, 1956, 30 min.
13. Sir William Haley, loc. cit.
14. See Spiro Agnew's complete speech in Marvin Barrett (ed.), *Survey of Broadcast Journalism, 1969–1970* (New York: Grosset & Dunlap, 1970), pp. 131–138.
15. See "A Letter from Aleksandr Solzhenitsyn," in Harrison E. Salisbury, *The Eloquence of Protest* (Boston: Houghton Mifflin Company, 1972), pp. 127–129.

16. Edward Jay Epstein, *News from Nowhere* (New York: Vintage Books, 1974), pp. xiv, xvii.
17. Ibid., p. 27 for alteration in story "play," p. 138 for small number of correspondents.
18. Ibid., p. 244. Also see the analysis on pp. 244–257.
19. William L. Rivers and Michael J. Nyhan, *Aspen Notebook on Government and the Media* (New York: Praeger Publishers, 1973), pp. 79–80, 83.
20. Ibid., p. 85.
21. See "America's News Industry," *U.S. News and World Report,* vol. 76, no. 17, Apr. 29, 1974. Pp. 32–35.
22. See Lester Markel, "Proper Tending of the Fourth Estate," *New York Times,* June 15, 1974.
23. Charles Renbar, "The First Amendment on Trial," *The Atlantic,* vol. 231, no. 4 (April 1973), pp. 45–54.
24. S. G. W. Benjamin, "A Group of Pre-Revolutionary Editors: Beginnings of Journalism in America," *Magazine of American History,* vol. 17, pp. 1–28, January–June 1887. For Mather statement, see Lucy Maynard Salmon, *The Newspaper and Authority* (New York: Oxford University Press, 1923), p. 75.
25. See Edwin Emery and Henry Ladd Smith, *The Press and America* (New York: Prentice-Hall, 1954), pp. 72–76. Also see *The Trial of John Peter Zenger of New York, Printer* (London: printed for J. Almon, 1765), pp. 11–12, 14–15, 44–45.
26. Speech of Mr. Livingston, June 21, 1798, *Annals of the Congress of the United States, Debates and Proceedings,* 5th Cong., May 15, 1797–Mar. 3, 1799 (Washington: Gales and Seaton, 1851), p. 2008. For Gallatin remarks, see p. 2162.
27. See *Debates in the House of Delegates of Virginia upon Certain Resolutions before the House, upon the Important Subject of the Acts of Congress Passed at Their Last Session, Commonly Called, the Alien and Sedition Laws* (Richmond, Va.: printed by Thomas Nicolson, 1798), p. 113.
28. See Albert Ellery Bergh (ed.), *The Writings of Thomas Jefferson* (Washington: The Thomas Jefferson Memorial Association, 1907), pp. 61–62.
29. See Henry Steele Commager, *Documents of American History* (New York: Appleton-Century-Crofts, 1948), pp. 179–182.
30. William G. Bleyer, *Main Currents in the History of American Journalism* (Boston: Houghton Mifflin Company, 1927), p. 121.
31. See Bernard Rubin, *Political Television* (Belmont, Calif.: Wadsworth Publishing Company, 1967), p. 51.
32. Ibid., p. 73.
33. Ben H. Bagdikian, *The Effete Conspiracy* (New York: Harper & Row, 1972), p. 106.
34. See Bernard Rubin, op. cit., pp. 86–88, 89–91.
35. Ibid., p. 100.

36. See Thomas Oliphant, "Watergate: The Case That Will Not Close," *Boston Globe,* May 5, 1974.
37. See John Herbers, "Nixon Innocence Affirmed in Brief," *New York Times,* May 1, 1974.
38. For background material see, for one example, Staff of the *Washington Post, The Fall of a President* (New York: Dell Publishing Co., 1974).
39. The networks may have been slow to gear up for the Watergate story, caught up as they were in coverage of traditional campaign business in 1972. Caution, bred from the Nixon administration's attacks on press objectivity since 1969, also played a part. On Oct. 27, 1972, however, Walter Cronkite, on the CBS Evening News, devoted the bulk of available time to the unfolding story. See Ben H. Bagdikian, "The Fruits of Agnewism," *Columbia Journalism Review,* vol. 11, no. 5, p. 20, January–February 1973.

CHAPTER 4

1. For political candidates and the Fairness Doctrine, see *Political Broadcast Catechism,* 7th ed. (Washington: National Association of Broadcasters, September 1972), pp. 1, 42. For right of reply to personal and political attacks, see *Red Lion Broadcasting Co., Inc., v. Federal Communications Commission; United States v. Radio Television News Directors Association,* 395 U.S. 367 (1969). For 1975 FCC ruling, see *News,* Report No. 13623, Federal Communications Commission, Washington, Sept. 28, 1975. The FCC declared that "the undue stifling of broadcast coverage of news events involving candidates for public office has been unfortunate." This action was responsive to questions raised in petitions from the Aspen Institute Program on Communications and Society and by CBS, Inc. Aspen Institute wanted exemptions under Section 315 for certain joint appearances of political candidates—primarily in debates—which it considered to be bona fide news events; CBS urged that Presidential news conferences be exempt when broadcast live. The FCC responded affirmatively to both petitions and broadened the press conference exemption to include governors, mayors, and other candidates whose press conferences may be considered newsworthy and suitable for on-the-spot coverage. There were two significant dissents: Commissioner Robert E. Lee observed: "What is clear to me is that the majority has sidestepped the very purpose of Section 315 . . . that all qualified candidates for public office be given equal opportunities to present their images and positions . . . via broadcast media." Commissioner Benjamin L. Hooks bluntly stated: "By exempting two popular forms of political weaponry, the press conference and the debate, the delicate balance of egalitarian precepts underlying political 'equal time' legislated into Section 315 and refined over 15 years of consistent administrative and judicial construction has suffered a severe and, perhaps, mortal blow."

2. Avery Leiserson, "Notes on the Theory of Political Opinion Formation," *The American Political Scicnce Review,* vol. 47, no. 1, p. 176, March 1953. The issues are fully explored in pp. 175–177.
3. See *In Re Applications of Alabama Educational Television Commission, Before the Federal Communications Commission,* FCC 74–1385, 29210; decision adopted Dec. 17, 1974, and released on Jan. 8, 1975; Federal Communications Commission, Washington, 39 pp. including text, appendix, and dissents. See especially pp. 2–4, 8–10, 25–27. As is the case in Alabama, South Carolina's black citizens constitute approximately 30 percent of the population. South Carolina's Educational Television (ETV) is seventeen years old and very generously funded by the state. It is well known for the high quality of its instructional programs. With its 1973 budget of over 6 million dollars, it reaches 650,000 students (one-third of the total) and more than 87,000 adults in special courses. Despite many positive distinctions separating this system from Alabama's in early 1975, the chairman of the state's Human Affairs Commission complained that only two programs were aimed at black people specifically. At that time, of ETV's 169 professional staff workers, only 19 were black. Also, after years of great achievements in other areas, "South Carolina's illiteracy rate today is topped only by Mississippi and Louisiana, and only Arkansas and Kentucky have a lower median educational level." See Jan Collins Stucker, "South Carolina's Educational Television Heads the Class," *New York Times,* Mar. 2, 1975, sec. 2.
4. See Jerome A. Barron, *Freedom of the Press for Whom?* (Bloomington, Ind.: Indiana University Press, 1973), pp. 270–303. Also see *Beauharnais v. Illinois* 343 U.S. 250 (1952).
5. See Peter Anderson, "How Media Handled School Busing Story," *Boston Sunday Globe,* Sept. 22, 1974.
6. George Creel, *How We Advertised America* (New York: Harper & Brothers, 1920), p. 156.
7. See *The Activities of the Committee on Public Information,* The Committee on Public Information, Washington, Jan. 27, 1918, pp. 3–6.
8. See *Preliminary Statement to the Press of the United States,* The Committee on Public Information, Washington, May 28, 1917, p. 6.
9. See Elmer Davis, "War Information," in Daniel Lerner (ed.), *Propaganda in War and Crisis* (New York: George W. Stewart, Publisher, 1951), p. 277. Italics added.
10. See Bernard Rubin, "Secrecy, Security and Traditions of Freedom of Information," in O. Lerbinger and A. Sullivan (eds.), *Information, Influence and Communication* (New York: Basic Books, 1965), p. 140.
11. Alfred E. Smith, *Progressive Democracy* (New York: Harcourt, Brace and Company, 1928), pp. 273–278.
12. See *Kent v. Dulles,* 35 7 U.S. 116 (1958).
13. See Peter Kihss, "A.P. Upholds Duty of U.S. and the Press to Report All Facts," *New York Times,* Apr. 22, 1958.

14. Joseph and Stewart Alsop, "That Washington Security Curtain," *The Saturday Evening Post,* vol. 227, no. 34, p. 32, Feb. 19, 1955.
15. See the Honorable John E. Moss, "Is There a 'Paper Curtain' in Washington?" Congressional Record, Extension of Remarks section, Aug. 2, 1955. See also U.S., Congresss, House, Committee on Government Operations, Special Subcommittee on Government Information, *Replies from Federal Agencies to Questionnaire,* 1955, pp. 5664, 368–371.
16. See U.S., Congress, House, Fifth Report of the Committee on Government Operations, *Availability of Information from Federal Departments and Agencies, House Report* no. 818, 1961, p. 7.
17. See Joseph and Stewart Alsop, "Matter of Fact: How the Censorship Works," *New York Herald Tribune,* June 13, 1955. Also see James Reston, "Officials and the Press," *New York Times,* Jan. 18, 1956.
18. See "Text of Eisenhower's Speech at Buffalo on His New York Campaign," *New York Times,* Oct. 24, 1952, sec. 4.
19. See David Halberstam, *The Best and the Brightest* (Greenwich, Conn.: Fawcett Publications, 1972), pp. 602–603.
20. Since documentation for these conclusions is more than ample, only illustrative key sources are cited. On aspects of the Vietnam war, see Francis Fitzgerald, *Fire in the Lake* (New York: Vintage Books, 1972); Dale Minor, *The Information War* (New York: Hawthorn Books, 1970); Thomas Powers, *The War at Home* (New York: Grossman Publishers, 1973); Joseph C. Goulden, *Truth Is the First Casualty* (Chicago: Rand McNally & Company, 1969); Adam Yarmolinsky, *The Military Establishment* (New York: Harper & Row, 1971); John E. Mueller, *War Presidents and Public Opinion* (New York: John Wiley & Sons, 1973); William McGaffin and Erwin Knoll, *Anything but the Truth* (New York: G. P. Putnam's Sons, 1968); David Halberstam, *The Best and the Brightest* (Greenwich, Conn.: Fawcett Publications, 1973 ed.). On aspects of recent Presidential relations with the press, see Staff of the *Washington Post, The Fall of a President* (New York: Dell Publishing Co., 1974); George E. Reedy, *The Twilight of the Presidency* (New York: New American Library, 1970); David S. Broder, *The Party's Over* (New York: Harper & Row, 1972); Carl Bernstein and Bob Woodward, *All the President's Men* (New York: Warner Paperback Library, 1974); Sanford J. Ungar, *The Papers and the Press* (New York: E. P. Dutton & Co., 1972); Neil Sheehan, et al., *The Pentagon Papers* (New York: Bantam Books, 1971).
21. See Arthur M. Schlesinger, Jr., *The Imperial Presidency* (New York: Popular Library, 1974), pp. 135, 157, 179.
22. See Lester Markel, *What You Don't Know Can Hurt You* (New York: Quadrangle/The New York Times Book Co., 1973), pp. 186–189.
23. See Norman E. Isaacs, "Beyond the 'Caldwell' Decision: 1," *Columbia Journalism Review,* vol. 11, no. 3, p. 19, Sept.–Oct. 1972.
24. See Fred P. Graham and Jack C. Landau, "The Federal Shield Law We Need," *Columbia Journalism Review,* vol. 11, no. 6, p. 27, March–April 1973.

25. Ibid., p. 26.
26. Norman E. Isaacs, op. cit., pp. 19–20.
27. See Don R. Pember, "The Pentagon Papers Decision: More Questions than Answers," *Journalism Quarterly,* vol. 48, no. 3, pp. 403–411, Autumn 1971.
28. See David K. Shipler, "Thirty Cases in Which Police or Courts Allegedly Threatened Free Press," *New York Times,* Feb. 18, 1973.
29. See Daniel Schorr, "Shadowing the Press," *The New Leader,* vol. 55, no. 4, pp. 8–10, Feb. 21, 1972. For more details on the story, see David Wise, "The President and the Press," *The Atlantic Monthly,* April 1973, pp. 55–64, especially pp. 61–63. Also see Julius Duscha, "The White House Watch over TV and the Press," *New York Times Magazine,* Aug. 20, 1972.
30. See Fred W. Friendly, "The Campaign to Politicize Broadcasting," *Columbia Journalism Review,* vol. 11, no. 6, pp. 9–18, especially p. 11, March–April 1973.
31. See U.S., Congress, Senate, Committee on the Judiciary, Subcommittee on Constitutional Rights, *Freedom of the Press: Hearings,* Sept. 28–30, Oct. 12–14, 19–20, 1971, and Feb. 1–2, 8, 16–17, 1972, pp. 80–81.
32. See U.S., Department of Justice, *Attorney General's Memorandum on the Public Information Section of the Administrative Procedure Act,* June 1967, pp. iii–iv. See pp. 30–39 for exemptions including, "The provisions of this section shall not be applicable to matters that are (1) specifically required by Executive Order to be kept secret in the interest of national defense or foreign policy." Also exempted were matters "related solely to the internal personnel rules and practices of any agency" and "matters that are . . . investigatory files compiled for law enforcement purposes except to the extent available by law to a private party."
33. Martin Arnold, "The New Rules on Freedom of Information," *New York Times,* Feb. 16, 1975. Also see Martin Arnold, "Congress, the Press and the Federal Agencies Are Taking Sides for Battle over Government's Right to Secrecy," *New York Times,* Nov. 15, 1974.
34. U.S., Congress, Senate, Committee on the Judiciary, Subcommittee on Administrative Practice and Procedure and Separation of Powers, *Freedom of Information, Executive Privilege, Secrecy in Government; Hearings,* vol. 2, June 7, 8, 11, 26, 1973, pp. 35, 43–44.
35. Ibid., *Hearings,* Apr. 10–12 and May 8–10, 16, 1973, vol. 1, pp. 262–263.
36. John E. Moss and Benny L. Kass, Esq., "The Spirit of Freedom of Information," *Trial Magazine,* March–April 1972, pp. 14–15. Also see Warren Weaver, Jr., "High Court Upsets News Report Curb," *New York Times,* Mar. 4, 1975. Even on the most sensitive issue of whether the news media can print or broadcast identities of rape victims, the U.S. Supreme Court recently ruled that the media cannot be enjoined from publication of "truthful information contained in court records." By a vote of 8 to 1 (Justice William Rehnquist dissented) the Justices ruled against a Georgia law that provided it was a mis-

demeanor to broadcast or print names of rape victims. In his dissent Justice Rehnquist noted that "the Court unquestionably places this issue on a par with the core First Amendment interest involved in protecting the press in its role of providing uninhibited political disclosure."

37. See Freedom of Information Act documentation in *Congressional Record,* Oct. 7, 1974, pp. H10001–H10014; Nov. 18, 1974, p. H10705; Nov. 20, 1974, pp. H10864–H10875; Nov. 21, 1974, pp. S19806–S19823.
38. For an important theoretical contribution, see Robert Axelrod, "Schema Theory: An Information Processing Model of Perception and Cognition," *The American Political Science Review,* vol. 67, no. 4, pp. 1248–1266, December 1973). For quotations, see pp. 1248–1249, 1265.
39. For the startling exposé of the story behind the Red Lion case of 1969 and further related matters during the Kennedy, Johnson, and Nixon administrations, see Fred W. Friendly, "What's Fair on the Air," *New York Times Magazine,* Mar. 30, 1975, pp. 11–13, 37–39, 41, 43.
40. See Les Brown, "Reported Political Use of Radio Fairness Doctrine under Kennedy and Johnson Is Causing Concern," *New York Times,* Mar. 31, 1975.
41. See Henry Weinstein, "Inquiry into Shell's Role on Broadcast Data Asked," *New York Times,* Sept. 13, 1974.
42. See advertisement of Mobil Oil Corporation, "Musings of an Oil Person . . . ," *Boston Evening Globe,* July 22, 1974, p. 17.
43. See Tom Wicker, "Freedom Is the Issue," *New York Times,* Sept. 29, 1974.
44. See Les Brown, "F.C.C. to Begin an Inquiry on Better TV for Jersey," *New York Times,* Feb. 1, 1975.
45. Hearings Before the Federal Communications Commission, *In re Complaint of Puerto Rican Media Action and Educational Council, Inc., et al. against Educational Broadcasting Corporation, Licensee of Station WNET-TV, Newark, New Jersey,* FCC 75–192, 30101, Feb. 28, 1975, 25 pp.
46. See *Miami Herald Publishing Company v. Tornillo,* 94 S. Ct. 2831 (1974). Also see Martin Arnold, "Florida's Press Law Raises Fundamental Questions," *New York Times,* June 14, 1974; Warren Weaver, "Rules for the Candidate under Attack Called Freedom Curb," *New York Times,* June 26, 1974. For other new aspects of the right of reply, vis-à-vis libelous attacks on citizens made by the press, see *Gertz v. Robert Welch, Inc.,* 94 S. Ct. 2297 (1974), in which the Court, by the narrowest margin of 5 to 4, decided the characteristics of a *private* person as distinct from a *public* person.

CHAPTER 5

1. Arthur H. White, "Election 1972: The Facts and Interpretations" (New York: Daniel Yankelovich, Inc., Nov. 29, 1972), mimeographed.
2. Mike Mansfield, "Impeachment on TV: In the Public Interest," *New York Times,* May 26, 1972. Also see Charles S. Steinberg, "Impeachment on

Radio," *New York Times,* Apr. 22, 1974. Steinberg (now professor of communications at Hunter College and formerly a vice president of the CBS television network) offered the interesting proposal that, if impeachment proceedings came about, the Senate's deliberations be covered by "national radio." Television would be restricted to news specials and reportage in the nightly news programs, to avoid a "theatrical extravaganza." According to Steinberg, "Despite its unparalleled power as a medium, television is volatile and ephemeral. It creates, despite its relentless verisimilitude, a pseudo environment."

3. See U.S., Senate, Select Committee on Presidential Campaign Activities, *The Final Report* (Washington, D.C.: U.S. Government Printing Office, 1974), pp. xxxi–xxxii.
4. Editorial, "A New Gilded Age," *New York Times,* Apr. 22, 1973.
5. See Staff of the *Washington Post, The Fall of a President* (New York: Dell Publishing Co., 1974), pp. 198–202.
6. See Ben H. Bagdikian, "The Fruits of Agnewism," *Columbia Journalism Review,* vol. 11, no. 5, p. 20, January–February 1973. Also, in the same issue, see Edwin Diamond, "Fairness and Balance in the Evening News," p. 23.
7. See *Current American Government* (Washington: Congressional Quarterly, Fall 1972), p. 49. Also see Albin Krebs, "TV Networks to Spend $9-Million Covering Election," *New York Times,* Nov. 7, 1972; Ben A. Franklin, "Campaign Spending in '72 Hit Record 400-Million," *New York Times,* Nov. 19, 1972. Krebs reported that Senator John G. Tower (Republican, Texas) spent more than 2.5 million dollars in his re-election campaign.
8. See "Federal Election Campaign Act of 1971: Conference Report," *Congressional Record,* vol. 117, no. 196, p. S21633, Dec. 14, 1971.
9. Bernard Rubin, "Political Television and the 1972 Election," *Polity,* vol. 6, no. 3, pp. 409–410, Spring 1974.
10. See "Text of Senator Nixon's Broadcast Explaining Supplementary Expense Fund," *New York Times,* Sept. 24, 1952.
11. See "Federal Election Campaign Amendments of 1974," Conference Report on S.3044, *Congressional Record,* vol. 120, no. 154, pp. H10326–10345, Oct. 10, 1974. For summary of the law's provisions included in analyses by the Center for Public Financing of Elections, see pp. H10328–H10329. For disclosure requirements applicable to private organizations, see p. H10330.
12. See "A Challenge to Election Law 'Reforms'," *Congressional Record,* vol. 121, no. 9; pp. S1106–1113, Jan. 28, 1975. Also, for background to election finance problems, see Alexander Heard, *The Costs of Democracy* (Chapel Hill, N.C.: The University of North Carolina Press, 1960); Herbert E. Alexander, *Political Financing* (Minneapolis: Burgess Publishing Company, 1972); Louise Overacker, *Presidential Campaign Funds* (Boston: Boston University Press, 1946); *Voters' Time: Report of the Twentieth Century Fund Commission on Campaign Costs in the Electronic Era* (New York: The Twentieth Century Fund, 1969); *Dollar Politics* (Washington: Congressional Quarterly, 1971);

Delmer D. Dunn, *Financing Presidential Campaigns* (Washington: The Brookings Institution, 1972).

13. John S. Saloma and Frederick H. Sontag, *Parties: The Real Opportunity for Effective Citizen Politics* (New York: Alfred A. Knopf, 1972), p. 307.
14. Joseph Napolitan, *The Election Game: And How to Win It* (New York: Doubleday & Company, 1972), p. 2.
15. Ibid., p. 65.
16. See Robert Agranoff (ed.), *The New Style in Election Campaigns* (Boston: Holbrook Press, 1972); Robert MacNeil, *The People Machine* (New York: Harper & Row, 1968), especially chap. 6.
17. Mary Adelaide Mendelson and David Hapgood, "The Political Economy of Nursing Homes," in *The Annals* entitled *Political Consequences of Aging,* vol. 415, p. 99, September 1974.
18. Timothy Crouse, *The Boys on the Bus* (New York: Ballantine Books, 1972), pp. 59, 153, 175.
19. David S. Broder, *The Party's Over: The Failure of Politics in America* (New York: Harper & Row, 1972), pp. 239, 251.
20. See Walter De Vries and V. Lance Tarrance, *The Ticket-Splitter: A New Force in American Politics* (Grand Rapids, Mich.: Wm. B. Eerdmans Publishing Co., 1972), pp. 15, 76, 92, 101–105, 125.
21. See Thomas E. Patterson and Robert D. McClure, *Political Advertising: Voter Reaction to Televised Political Commercials* (Princeton, N.J.: Citizens' Research Foundation, 1973), p. 35.
22. See Christopher Lydon, "Packaging Voters for Candidates, TV-Style," *New York Times,* Oct. 29, 1974.
23. Samuel J. Archibald, "A Brief History of Dirty Politics," in Ray Hiebert, Robert Jones, et al. (eds.), *The Political Image Merchants* (Washington: Acropolis Books, 1971), pp. 232–234.
24. David Broder, "And What about a Hitler?," in ibid., p. 22.
25. See Bernard Rubin, *Political Television* (Belmont, Calif.: Wadsworth Publishing Company, 1967), pp. 192–194.

INDEX